786-110

IGCSE
CONCEPTUAL CHEMISTRY

Volume-1

As per latest IGCSE syllabus 2017-19

Husain Pipliyawala

Copyright © 2016 Husain Pipliyawala

All rights reserved.

ISBN: 152345301X
ISBN-13: 978-1523453016

DEDICATION

With kind benedictions of His Holiness Dr. Sayyedna Mufaddal Saifuddin (TUS) and due respect, I dedicate this book to my mother Ms. Farida, family and my chemistry teachers Prof. Anil Chincholikar and Prof. Pankaj Chandorkar without the basic knowledge and teaching's from them my chemistry understanding would not of this level to author this book for world class IGCSE curricula provided by prestigious and renowned boards and universities. Also the credit goes to my students who asked me several questions each lecture I took, that made me realize what a book should have for a student to have interest in a technical subject like IGCSE Chemistry.

CONTENTS

Acknowledgments viii

Chapter-1 PARTICULATE NATURE OF MATTER   Page-1

- Matter 1
- Arrangement of particles in matter 1
- Kinetic Particle Theory 1
- Conversion of States 2
- Comparison of three states 2
- Interconversion of states 3
- Heating Curve 4
- Cooling Curve 5
- Brownian Motion 6
- Exam Style Questions 7

Chapter-2 MEASUREMENT, EXPERIMENTAL TECHNIQUES   Page 12

- Measurement 12
- Apparatus for Measuring 12
- Criteria for Purity 13
- Difference between compounds and mixtures 14
- Heterogeneous mixtures 15
- Homogenous mixtures 15
- Separation Techniques 15
- Decantation 15
- Chromatography 16
- Filtration 17

- ➤ Sublimation — 18
- ➤ Distillation — 18
- ➤ Fractional distillation — 19
- ➤ Centrifugation — 21
- ➤ Evaporation — 21
- ➤ Crystallization — 22
- ➤ Exam Style Questions — 23

Chapter-3 STRUCTURE OF ATOM — Page 34

- ➤ Atoms — 34
- ➤ Daltons Atomic Theory — 34
- ➤ Elements — 34
- ➤ Discovery of sub atomic particles — 34
- ➤ Electron — 35
- ➤ Proton — 35
- ➤ Gold Foil Experiment — 35
- ➤ Neutron — 36
- ➤ Isotopes — 37
- ➤ Radio Isotopes — 37
- ➤ Electronic Arrangement — 38
- ➤ Exam Style Questions — 40

Chapter-4 PERIODIC TABLE — Page 47

- ➤ Dobernier's Law of triads — 47
- ➤ Groups and Periods — 48
- ➤ Metals and Non-metals — 48
- ➤ Alkali Metals — 50
- ➤ Alkaline Earth Metals — 51

- ➢ Halogens — 52
- ➢ Transition elements — 53
- ➢ Noble gases — 54
- ➢ Exam Style Questions — 55

Chapter-5 STOICHIOMETRY /MOLE CALCULATION   Page 62

- ➢ Elements — 62
- ➢ Compounds — 62
- ➢ Chemical Formula — 62
- ➢ Word Equation — 63
- ➢ Symbol Equation — 63
- ➢ Rules for Balancing Chemical Equations — 63
- ➢ Relative Atomic Mass — 64
- ➢ Naming Compound — 65
- ➢ Information from a chemical equation — 66
- ➢ Definition of Mole — 66
- ➢ The mole concept — 66
- ➢ Molar Mass — 67
- ➢ Important Formulae — 67
- ➢ Limiting reagent — 68
- ➢ Reacting masses and ratios — 68
- ➢ Molar Volumes — 70
- ➢ Concentration of Solutions — 70
- ➢ Water of Crystallization — 71
- ➢ Empirical and Molecular Formula Percentage Yield — 72
- ➢ Percentage Yield — 73
- ➢ Percentage Purity — 74
- ➢ Concept Practice Questions — 75
- ➢ Exam Style Questions — 78

Chapter-6 CHEMICAL BONDING   Page 90

- Chemical Bond   90
- Ionic Bond   90
- Lattice   92
- Covalent Bond   93
- Giant Covalent Structures   94
- Diamond   95
- Graphite   95
- Silica   96
- Metallic Bond   97
- Coordinate Bond   98
- Valencies of Common ions   98
- How to write a Chemical Formula   99
- Exam Style Questions   101

ACKNOWLEDGMENTS

"I would like to express my sincere gratitude to my wife Sakina and daughter Amatullah who saw me through this book with each and every piece of work and detail without disturbing me ; to Mr. Subrahmanyam Atchula for encouragement and suggestions for layout.

Also would like to thank Amazon for enabling me to publish this book. Above all I want to thank rest of my family, who encouraged and supported me in spite of all the time it took me away from them. It was a long and difficult journey for them.

Lastly, I request forgiveness of all those who have been with me over the course of the years in my learning journey of chemistry and whose names I have failed to mention."

CHAPTER-1

PARTICULATE NATURE OF MATTER

Matter: All materials around us constitute the matter. It is defined as anything that has mass and occupies space is called matter. Commonly there are three states of matter solid, liquid and gas. However, research has shown that matter can also exist as Plasma state and Bose Einstein Condensate as shown by Bose and Einstein. Below is the detailed study of three common states.

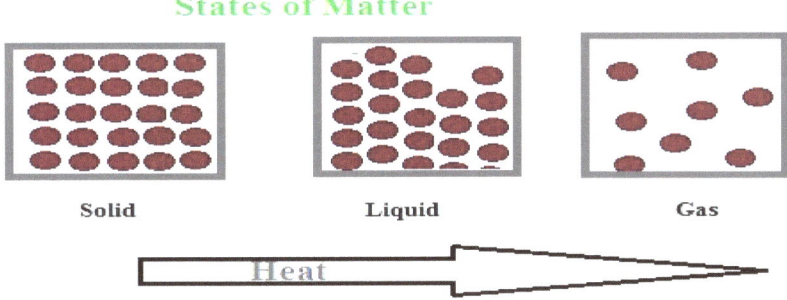

Kinetic Particle Theory

This theory explains the behavior of matter in different states of matter such as solid liquid and gas.

-According to this theory, particle in solids are compactly packed and those in liquids and gases are loosely packed.

-These particles are in continuous random motion due to their energies (kinetic energy) especially particles of liquids and gases. This random motion of particles is called diffusion.

-When matter is heated the kinetic energy of particles increases due to which they diffuse at a faster rate.

Conversion of States

Melting: It is the process of conversion of solid into liquid without any change in its temperature. For e.g. melting of ice

Boiling: The process of conversion of liquid to gas at boiling point without any change in its temperature. For e.g. Boiling of water. It is a bulk phenomenon as the heat gets distributed to the entire liquid.

Sublimation: The process of conversion of solid directly into gas on heating is called sublimation.

Deposition: The process of phase change of a gas directly to solid. For example water vapor deposition on windscreens of vehicles in winter.

Property	Solid	Liquid	Gas
Particular arrangement	Compactly arranged in a fixed position	Loosely arranged and are free to move	Very loosely packed and are free to move
Particular movement	Particles vibrate about fixed position	Particles randomly move in layers	Particles move randomly in all directions
Compressibility	Not compressible	Compressible to some extent	Highly compressible
Shape and Appearance	Have fixed shaped and volume	Have no fixed shape but fixed volume	Have no fixed shape or volume
Forces of attraction between the particles	Strong forces of attraction are present	Weak forces of attraction are present	No attractive forces

Interconversion of States

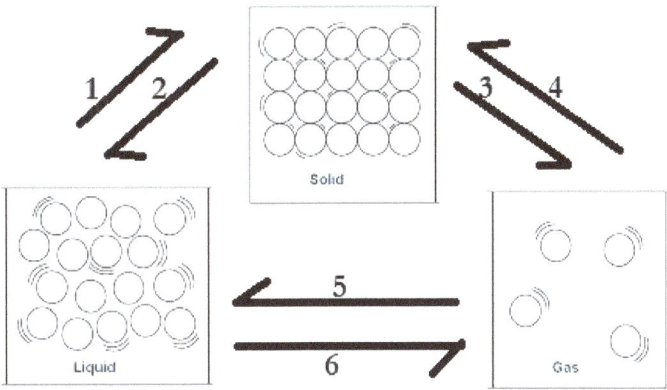

1 = Freezing   2 = Melting   3 = Sublimation

4 = Deposition   5 = Condensation   6 = Boiling

Freezing: The change of state from liquid to solid at melting point is called freezing. For e.g. Formation of ice from water

Condensation: It is the conversion of gas to liquid without any change in temperature. For e.g. Formation of water droplets from steam.

Sublimation: The conversion of a solid directly into gas is called sublimation.

Evaporation: The change of state from liquid to gas at any temperature (below boiling point) is called evaporation. It is a surface phenomenon.

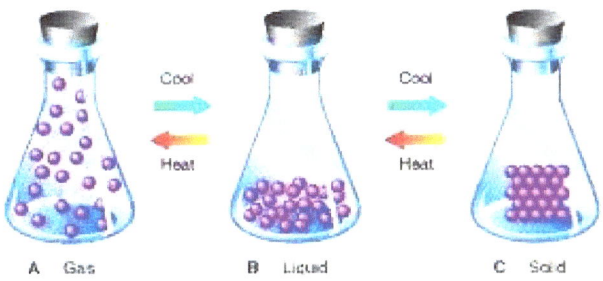

Heating and Cooling Curves:

These are the graphs that explain the change in states of a substance with temperature and time. A heating curve is a graph plotted between temperature and time for heating a given sample (generally solid). For e.g. heating curve of a substance is shown below where T-1 is the melting point and T-2 is the boiling point of the substance.

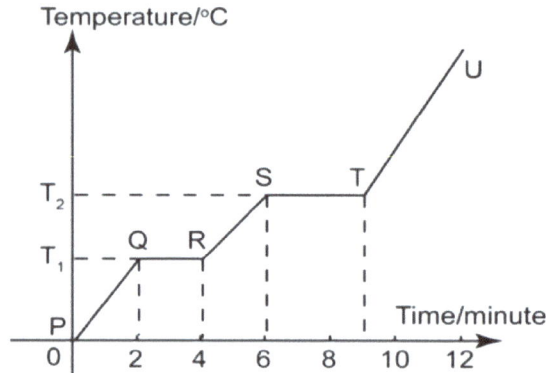

Difference between evaporation and boiling is that evaporation is conversion of upper surface of liquid into gas where is boiling happens in the bulk of the liquid. Also it is worth noting that evaporation is a surface phenomenon whereas boiling is a bulk phenomenon.

In separation, evaporation is removing the liquid from a solution, usually to leave solid crystals. It can be done quickly with gentle heating or left out to 'dry up' slowly in 'open air'. The solid will almost certainly be less volatile than the solvent and will remain as a crystalline residue. Evaporation is often followed by crystallization.

Important Points to Note:

1- During melting of a substance the temperature remains constant (melting point)

2- During boiling of a substance the temperature remains constant (boiling point).

3- The heat supplied during melting and boiling of a substance is used in breaking the inter molecular attractions therefore no change in temperature is recorded at these instances.

4- At slopes of the graph both the states exist together. For e.g. at RS slope solid and liquid exist together.

Cooling Curve: It is a graph that shows change in state of a substance from gas to lower energy states with temperature and time.

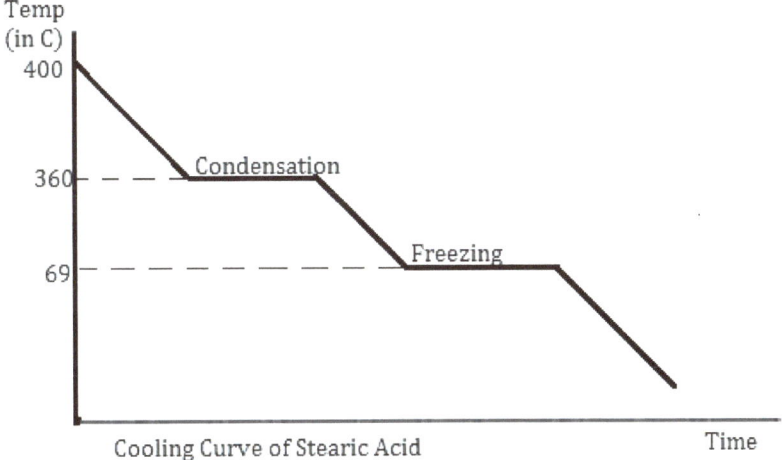

Cooling Curve of Stearic Acid

Important Points to Note:

1- During condensation of a gaseous substance to liquid the temperature remains constant (dew point).

2- During freezing of a liquid substance to solid the temperature remains constant (freezing point).

3- The heat released during dew point and freezing point of a substance is due to the making of inter molecular attractions therefore no change in temperature is recorded at these instances.

Brownian Motion:

Brownian motion got its name after the famous botanist Robert Brown, who first observed this motion in 1827 in a sample of pollen grains. He used a microscope to look at pollen grains that were moving randomly in water. It is defined as the movement of large particles in a fluid due to constant collision of these particles with small fluid particles. However Albert Einstein later explained this that the pollen grains were being moved by collision of water molecules. This confirmed the existence of atoms and molecules and provided evidence for kinetic particle theory.

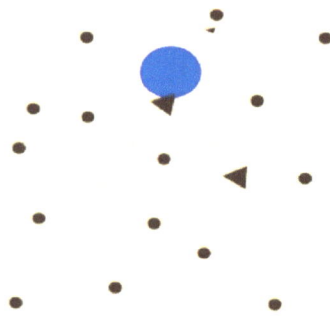

Brownian movement of pollen grain in water

Important Terminology

1. Freezing

2. Boiling

3. Condensation

4. Evaporation

5. Diffusion

Exam Style Questions :

Section A-Multiple Choice Questions

1. Some IGCSE students are asked to describe differences between liquids and solids. Two of their suggestions are:

1 Solids are highly compressible,

2 Liquids are somewhat compressible

Which suggestions is/are correct?

A 1 only

B 2 only

C both

D none

2. Three identical balloons are filled with different gases all at the same temperature and pressure.

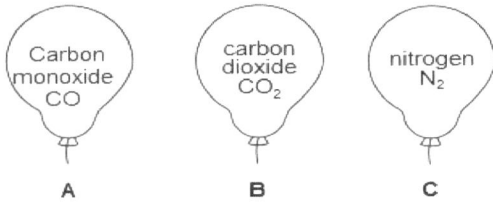

The gases gradually diffuse out of the balloons.

Which of balloons will deflate at the same rate?

A A and B   **B** B and C   **C** C and A   **D** A,B and C

3. The below diagram shows how to obtain pure water from seawater for drinking.

Where do water molecules have minimum energy?

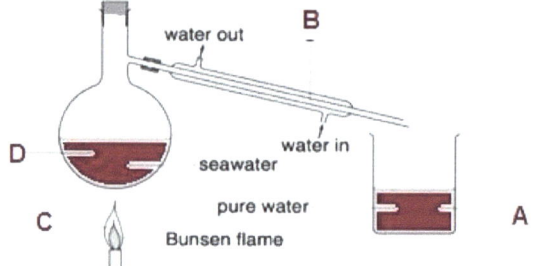

4. A solid substance is heated until it forms vapor. The graph shows the temperature of the solid during this process. Which part of the graph shows both liquid and gas states ?

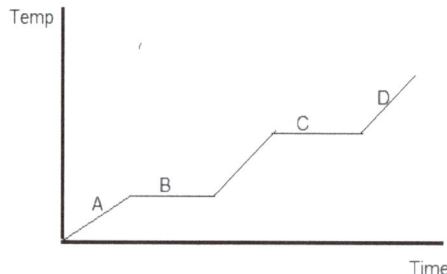

5. In which of the following are the particles arranged in a regular pattern?

A CO_2

B H_2O

C CCl_4

D CaO

Section B *Structured Questions*

1. Sulphur dioxide (SO_2) is a gas which forms an acidic solution when dissolved in water. Complete the diagram to show the arrangement of molecules in SO_2 gas.

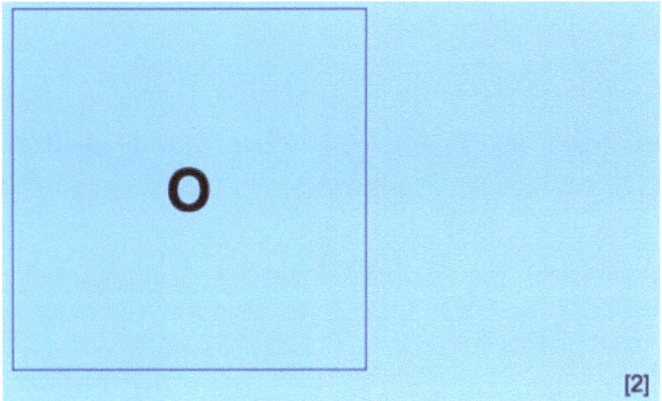

[2]

(b) Which one of the following values is most likely to represent the pH of a dilute solution of acetic acid? Put a ring around the correct answer.

pH1   pH5   pH12   pH9

(c) A solution of SO_2 has a strong smell? A beaker of SO_2 solution is put in the corner of a room.

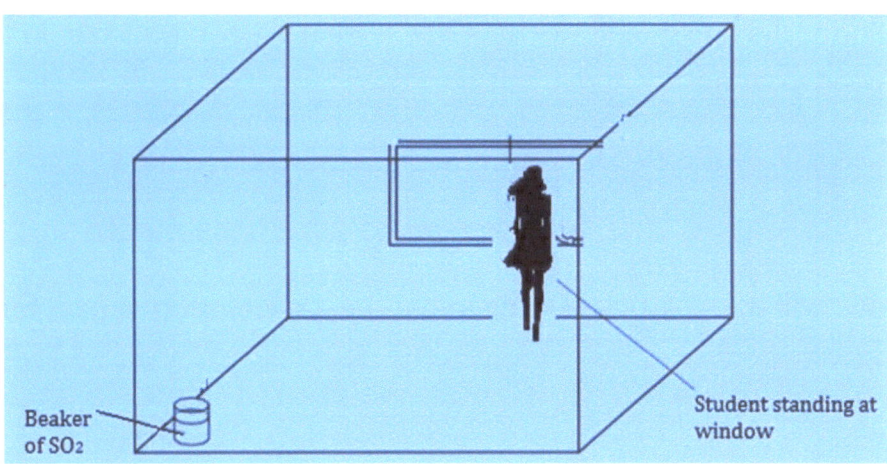

At first the student smells no SO_2. But after sometime she is able to smell Sulphur dioxide.

i-Explain in context with kinetic particle theory.

……………………………………………………………………………………

....

....

ii) The smell further becomes prominent when fan is switched on. Explain.

....

....

....

iii) How is arrangement of particles in Sulphur dioxide different in gaseous and liquid states.

....

....

....

iv) Sulphur dioxide condenses at -72^0C. What do you understand by the term condensation.

....

....

v) How will the particle arrangement be different in liquid Sulphur dioxide.

....

....

2. The diagram on the next page shows a biogas digester. Bacteria ferment animal and vegetable waste. The gas produced is a mixture of mainly hydrogen sulphide and methane and ethane.

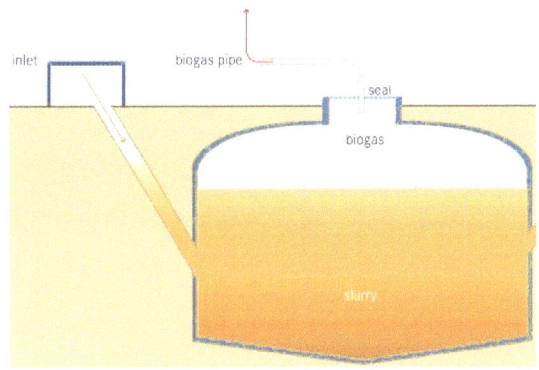

a) Describe the arrangement and motion of the molecules in ethane gas.

………………………………………………………………………………..

………………………………………………………………………………

………………………………………………………………………………

b) The biogas outlet is connected on the top of the digester. Why?

………………………………………………………………………………..

………………………………………………………………………………

………………………………………………………………………………

c) Methane has a low boiling point. What do you understand by boiling.

………………………………………………………………………………..

………………………………………………………………………………

d) How is boiling different from evaporation.

………………………………………………………………………………..

………………………………………………………………………………

………………………………………………………………………………

CHAPTER -2

EXPERIMENTAL TECHNIQUES

Measurement: It is the assigning a value in Number and unit to the characteristic of an object or event which is comparable to a standard attribute. For example measurement of volume can be done as 20cm3. Following are the common apparatus used for measurement during laboratory testing.

Apparatus for Measuring

Measuring cylinder: It is an instrument used to measure volumes of liquids in experiments quickly. If water or aqueous solutions are measured than lower meniscus in cylinder is noted.

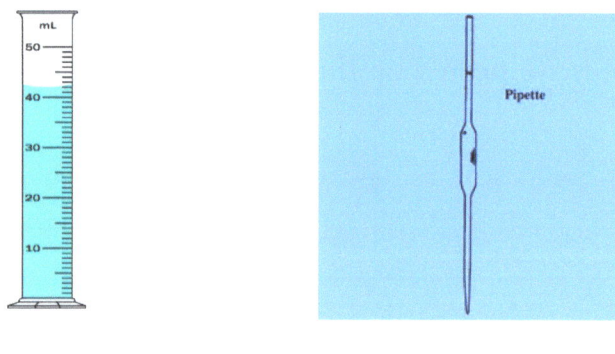

Measuring cylinder   *Pipette*

Pipette: It is an apparatus used to measure accurate volumes of liquids during experiments. A pipette is generally used in titrations for transferring liquids to the conical flask.

Note: Candidates are expected to name appropriate apparatus for the measurement of time, temperature, mass and volume, including burettes, pipettes and measuring cylinders.

Thermometer: It is an instrument used to measure temperature in laboratory and clinics. Lab thermometers commonly used contain mercury (silvery in appearance) alcohol (red in appearance). The measuring range of lab thermometer is from -10^0C to 110^0C.

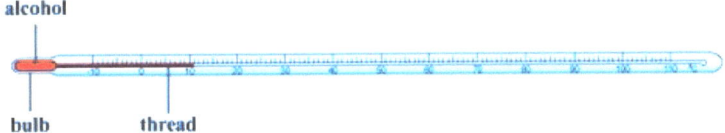

Burette : It is an instrument used for transferring accurate amount of measured volume of a liquid in an experiment up to 0.1cm3. Burette is used in titrations for noting down the amount of liquid transferred.

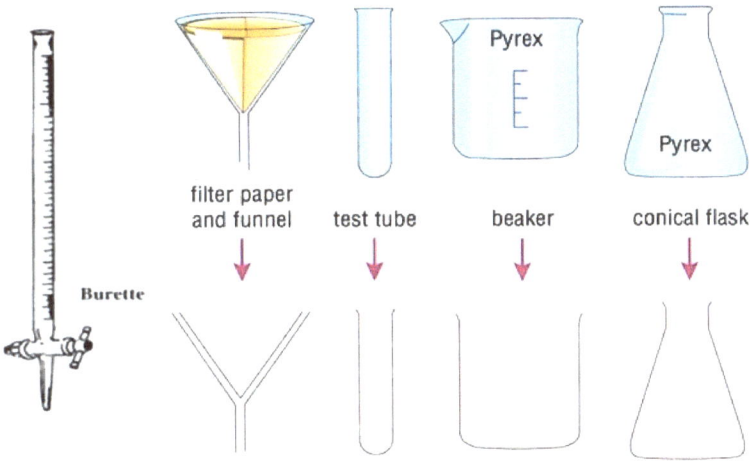

Above diagram shows a burette for accurate volume measurement and some common lab apparatus used for unmeasured amount of substance transferred.

Criteria of purity : Pure substance generally an element or a compound can be identified by sharp (fixed) melting and boiling point. For example, pure water freezes at 0^0C and boils at 100^0C. However, when impurities are present in a substance than its melting and boiling points are not fixed rather they have a range. When impurities are present in a substance than its melting point decreases. Similarly, presence of impurities also causes an increase in boiling point of an impure substance often referred as mixtures.

Mixtures : These are formed by physical combination of two or more substances and therefore have more than one type of atoms in them. The atoms of different elements are in an indefinite ratio in mixtures.

Comparative study of compounds and mixtures is given below.

Compounds	Mixtures
These are made up of two or more different atoms chemically combined.	These are made up of two or more different atoms physically mixed.
The atoms of different elements are in fixed ratio.	Atoms of different elements are in indefinite ratio.
They have characteristic chemical properties.	They have varying properties.
They have properties different from constituent's elements.	They have properties like their constituent elements.
For example, water is a compound of hydrogen and oxygen.	A container filled with samples of hydrogen and oxygen gas is a mixture.

(a) Mixture   (b) Element   (c) Compound

Types of mixtures- Mixtures are of two types –

i- Heterogeneous

ii- Homogenous

Heterogeneous mixtures: These are the mixtures that have non uniform composition throughout. For example, sand and water form a heterogeneous mixture as shown.

Heterogeneous Mixture of sand, mud and water. Another example can be fruit drink with pulp.

Homogeneous mixtures: These are the mixtures that have a uniform composition throughout. For example, mixture of salt dissolved in water.

Homogenous Mixture (solution) of salt and water appears clear just like pure water. All solutions are examples of this type

Separation Techniques: Since in daily life most of the substance contain impurities lab techniques can be used to purify the substance or obtain a specific component from the mixture. Following are some important separation techniques.

i) Decantation: It is the most common technique to separate insoluble components of a suspension. The mixture is allowed to stand undisturbed for some time and then the liquid is gently poured off.

ii) Chromatography: It is a technique used to separate a mixture of colored substances on the basis of difference in their solubility's in a given solvent. For separation of components of ink and dyes chromatography can be used. The technique is performed by using a special paper called chromatography paper however for a general separation filter paper can also be used.

• The technique is performed by drawing a pencil line on a chromatography paper and putting a thick spot of mixture on the center of line.

• This paper is than placed stationary for some time in a solvent such as water, alcohol or acetone such that the spot is just above the solvent front.

• The colored mixture than rises up the paper along with the solvent front due to its solubility in the solvent.

• The components of mixture such as colors of ink get separated and form separate spots on the paper that are easily identified.

• For a mixture that is colorless such as mixture of amino acids the spots are obtained by spraying the chromatogram with a locating reagent such as *ninhydrin*.

• For further comparisons and identification Rf (retardation factor) value of each spot is calculated by following formula

$$\text{Rf Value} = \frac{\text{Distance travelled by the spot}}{\text{Distance travelled by the solvent}}$$

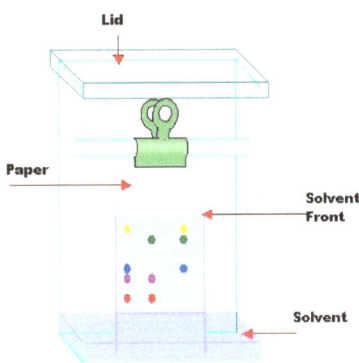

Uses of chromatography:

1. in separating components of ink and dyes.

2. in separating pigments of chlorophyll from leaves.

3. in separating amino acids from a mixture of amino acids.

iii) Filtration: This method is used to obtain insoluble solid from a liquid and requires the use of a filter paper and a funnel. The insoluble solid obtained in the filter paper are called residue and the liquid collected after filtration is called filtrate. For examples precipitates of salts such as lead sulphide can be obtained from a suspension by this method.

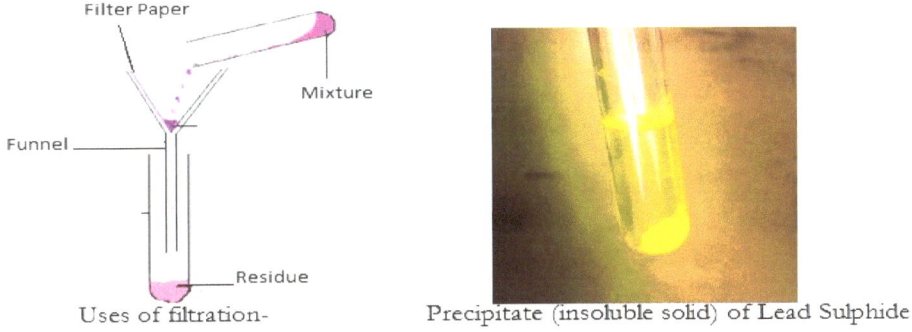

Uses of filtration-   Precipitate (insoluble solid) of Lead Sulphide

1. To separate tea leaves from tea.

2. To separate precipitate from a solution.

3. In separating sand from water.

iv) Sublimation: This technique is used to separate a mixture of solids such that one of them directly gets converted to gas (sublimes) on heating. For example a mixture of sodium chloride and ammonium chloride can be separated by sublimation. Other substances that can sublime on heating are camphor, dry ice etc.

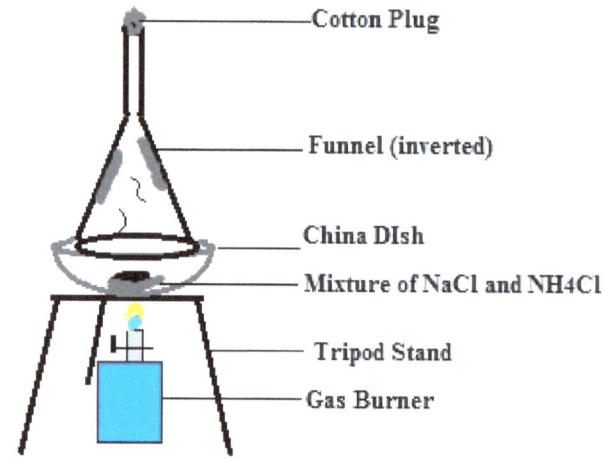

Laboratory Setup for Sublimation

v) Distillation : It is a technique to separate a mixture containing miscible (soluble) components generally two liquids or a solid dissolved in liquid. The principle of distillation is to separate the components on the basis of difference in their boiling points.

The apparatus consists of a round bottom flask containing the mixture to be separated. This flask is covered with a rubber bung and a thermometer to note temperature is place over it. Now heat is supplied to flask such that the fraction with low boiling point is obtained first by boiling followed by condensation. This process is performed in lab by using the following setup shown on the next page.

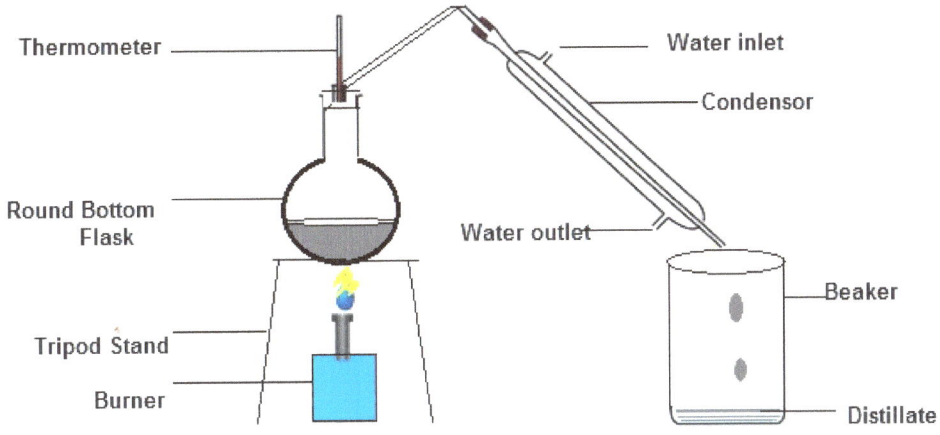

The component with higher boiling point still remains in the round bottom flask. The liquid obtained in the other flask is called distillate.

For example, the mixture of salt dissolved in water can be separated by distillation. Similarly, a mixture of water and alcohol can be separated by distillation. And also a mixture of organic liquids can also be separated by this method.

Uses of Distillation:

1. In separating water from solution of salt or sugar.

2. In separating alcohol from water.

vi) Fractional Distillation: It is a technique to separate a mixture containing more than two miscible (soluble) liquids on the basis of difference in their boiling points. For example, petroleum is the mixture of many hydrocarbons and can be separated by fractional distillation. The laboratory apparatus used in fractional distillation is almost same a simple distillation except a glass bead column is used in fractional distillation as shown. This column allows the vapor of higher boiling liquids to cool down and remain as liquid in a round bottom flask.

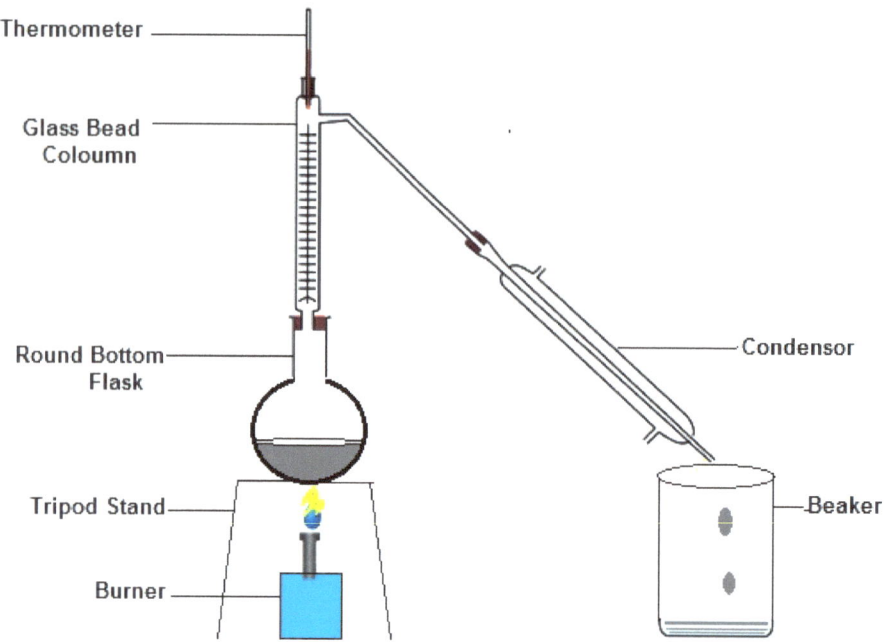

Fractional Distillation in Industries: In petroleum industry fractional distillation is performed in a fractionating column. It is huge tower that consists of bubble cap arrangements for fractions of petroleum to vaporize and the lightest fraction is collected at the top whereas the heaviest is collected at the bottom. The tower has outlets to collect the fractions at different levels. At each level the temperature is maintained according to the boiling point of the fraction.

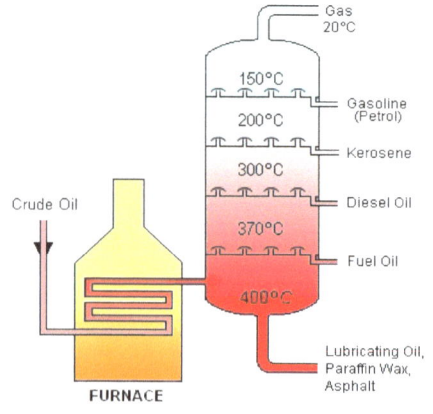

Fractionating Column for refining of petroleum in industry

Uses of Fractional distillation:

1. In separating fractions of petroleum

2. In separating mixture of alcohols

vii) Centrifugation: This process is used in separating a small amount of dissolved solid from a liquid. The principle of the process is difference in densities of the solid and liquid. When the mixture is spun very fast the solid being heavier hurls to the bottom of the beaker whereas the liquid remains in the top. This process is done in laboratory by an instrument called centrifuge.

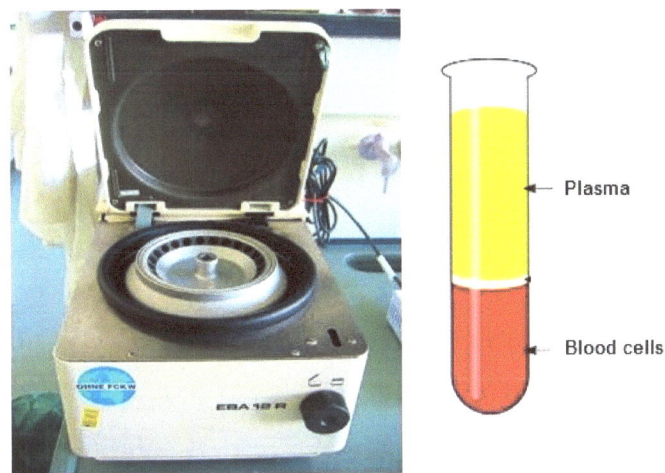

In a centrifuge samples of blood are placed to separate plasma from blood cells for pathological testing. The figure above shows a centrifuged sample of blood removed from a centrifuge.

Uses of centrifugation:

1. In drying of clothes in washing machine.

2. In separating blood cells from plasma.

viii) Evaporation: The process is used in obtaining a dry sample of dissolved solid from a solution (homogenous mixture of solid and liquid). The process is based on heating a solution till all the liquid gets vaporized and dry crystals of solid are obtained.

Uses of Evaporation :

1. Salt is obtained from seawater by evaporation.

2. Crystals of anhydrous salts such as KCl can be obtained by evaporating their aqueous solution by evaporation.

3. Evaporation is widely used in drying of clothes and wet floors after cleaning.

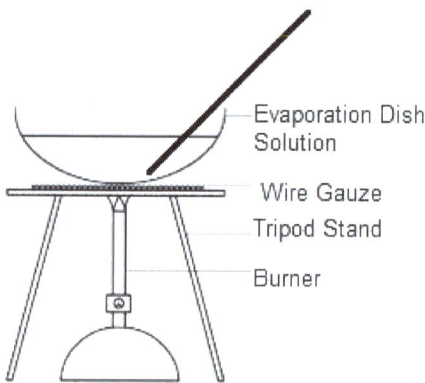

ix) Crystallization: This process is used to obtain crystals of soluble salts from their aqueous solutions. When a saturated solution of soluble salt such as copper sulphate is heated at crystallization point than crystals are formed on the wall china dish. Formation of crystals can be checked by a glass rod.

Uses of crystallization:

1. In obtaining crystals of soluble salts from solutions.

2. In obtaining crystals of hydrated salts from solutions.

Exam Style Questions

Section A MCQ (Multiple Choice Questions)

1. A student investigates if, at 35^0 C, the concentration of acid affects how rapidly it reacts with a known mass of calcium. The student has a beaker, concentrated acid, water and the apparatus below.

M a clock

N a balance

O a measuring cylinder

P a thermometer

Which of these pieces of apparatus does the student use?

A M, N and O only

B M, N and P only

C N, O and P only

D M, N, O and P

2. The boiling point of liquid Q is lower than that of water. To test a student, a teacher covers up the numbers on a alcohol thermometer. The student places the thermometer in boiling liquid Q. The diagram represents part of the stem of this thermometer.

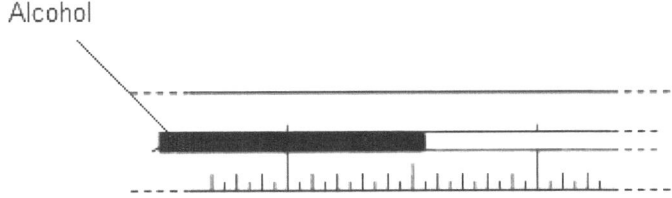

What could the temperature on the thermometer be?

A 65.5°C

B 82.5°C

C 104.5°C

D 105.5°C

3. Which of the following mixtures can be separated by adding water, stirring and filtering?

A potassium chloride and sodium chloride solutions

B copper and iron fillings

C diamond and graphite

D sand and powdered sodium nitrate

4. A student experimenting with magnesium carbonate powder is asked to measure the time taken for 2.50 g of magnesium carbonate to react completely with 50.0cm³ (an excess) of dilute hydrochloric acid. Which pieces of apparatus does the student need?

A balance, clock, pipette

B balance, clock, thermometer

C balance, pipette, thermometer

D clock, pipette, thermometer

5. Nickel(II)chloride crystals(soluble) are separated from sand using the four processes listed below. In which order are these processes used?

	Step-1	Step-2	Step-3	Step-4
A	filtering	dissolving	crystallising	evaporating
B	filtering	dissolving	evaporating	crystallising
C	dissolving	evaporating	filtering	crystallising
D	dissolving	filtering	evaporating	crystallising

6. Solid Y melts at exactly 54°C and boils at exactly 302°C. Solid Z, when dissolved in water and examined using paper chromatography, shows a blue colour and a red colour. Which row is correct?

	contains only one substance	contains more than one substance
A	Y and Z	–
B	Y	Z
C	Z	Y
D	–	Y and Z

7. The below diagram shows an experiment with aqueous ammonium chloride.

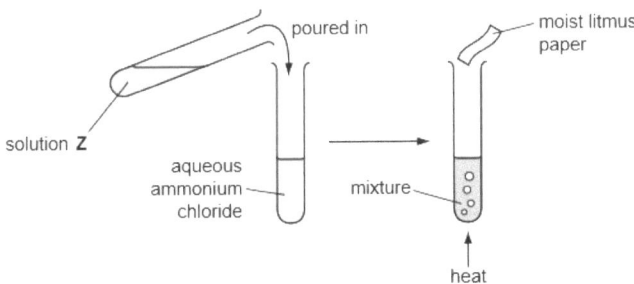

If the moist litmus turns blue what can be solution Z.

A Al(OH)$_3$

B HCl

C NaOH

D Distilled water

8. A student is asked to separate a suspension of sand in a solution of sodium chloride and ammonium chloride so as to obtain all three substances. He used the following processes. What do you think is the correct order to obtain all the three constituents?

A. sublimation; filtration; evaporation

B. sublimation; evaporation; filtration

C. filtration; evaporation; sublimation

D. evaporation; sublimation; filtration

9. The diagram shows a chromatogram of several inks.

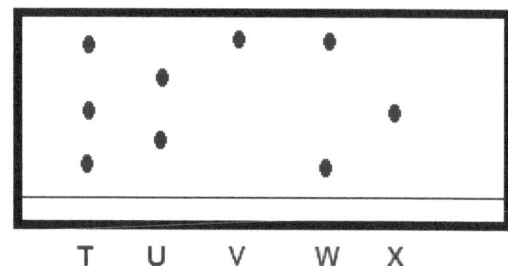

Which statement is correct?

A ink T is a combination of U and V

B ink T can be made by mixing ink W and X

C ink T can be made by mixing ink U and W

D ink T is a pure substance.

10. A mixture of solid and liquid was separated by centrifugation as shown. For experiment to be successful.

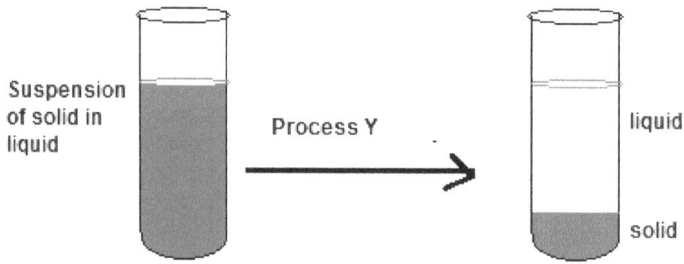

A. The mixture should be heated strongly

B. The solid should be soluble in liquid.

C. The solid should be lighter than liquid.

D. The liquid should be lighter than solid.

11. Sulphur dioxide is soluble in water, whereas chlorine is only slightly soluble in water. Both gases can be dried using concentrated sulfuric acid.

Which diagram represents the correct method of obtaining pure dry chlorine from damp chlorine containing a small amount of Sulphur dioxide?

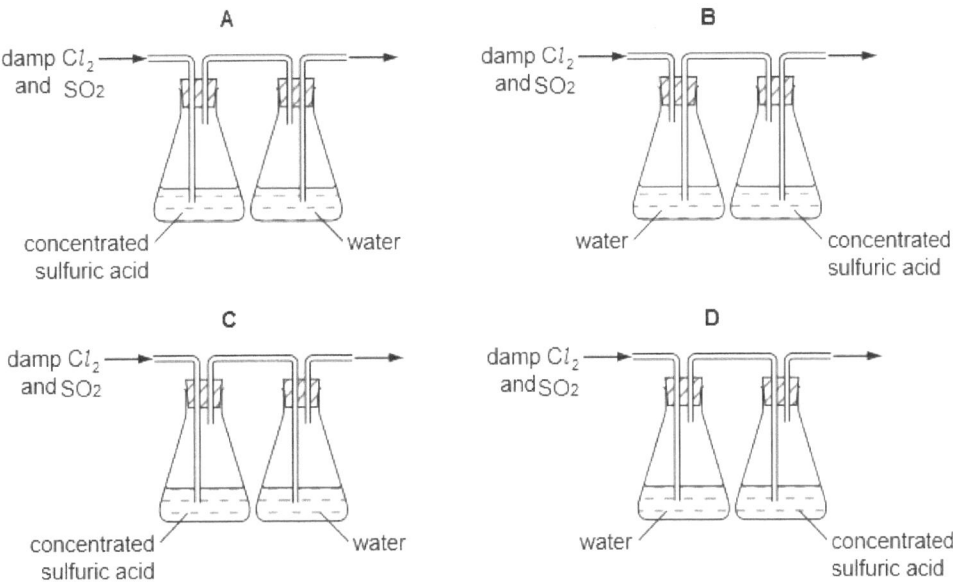

Section - B: Structured Questions

1. This diagram shows a fractionating column for the separation of petroleum in industry.

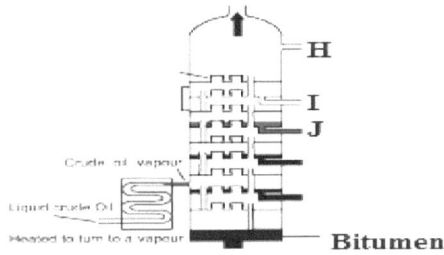

The following fractions leave the column.

fraction	boiling range / °C
Bitumen	above 700
paraffin	150 – 240
Diesel	220 – 250

(a) Which fractions leave the column at each of the points I and J ?

....

....

(b) Explain the process by which fractionating column separates the crude oil mixture.

....

....

....

(c) How the above process is different from simple distillation.

....

....

2. A sample of mixed fruit jam was investigated to check the four colorings present.

The mixed fruit jam was first boiled with water. The mixture was then filtered in second step. The filtrate was then concentrated in third step. Finally the concentrate was analyzed by chromatography paper.

(a) Why was the jam boiled with water?

....

....

(b) How was concentration performed?

....

....

(c) Draw a diagram to show the possible paper chromatogram obtained finally.

(d) Write one safety measure in this experiment.

....

(e) What was the purpose of filtration here?

....

....

3. A mixture of sodium chloride and ammonium chloride can be separated by sublimation. The apparatus below can be used to carry out such a separation in the laboratory.

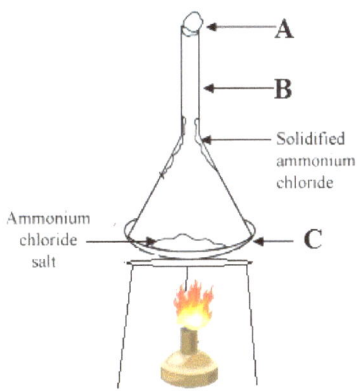

a) Name each piece of apparatus

A ....

B ....

C ....

b) What is the purpose of A?

....

c) How could the purity of the ammonium chloride collected be checked?

....

4. (a) Name the apparatus below and write one advantage it has over measuring cylinder.

....

....

b) How is pipette different from this apparatus.

....

....

c) (i) What safety item should be used with a pipette?

....

(i) Why is this safety item used?

....

5. The hydroxides of the Alkali metals are soluble in water. Crystals of potassium sulphate can be prepared from potassium hydroxide and sulphuric acid by titration.

burette filled with acid

conical flask

25.0 cm³ of aqueous KOH and indicator

a-i) A neutral solution of potassium hydroxide is obtained in the above experiment. What is the purpose of using indicator in the experiment.

....

....

ii) Why is it better to add acid slowly and mixing each time rather than running the burette tap continuously.

....

b) The acid used above is diluted by water before use, by adding acid to water. Why is it not advised to add water to acid for dilution and always to add acid to water.

....

....

6. An experiment was carried out using the gas syringe apparatus to collect the oxygen gas. Five trials were performed such that in each experimental trial the amount of reactants were doubled. The gas collected is shown by following syringe diagrams.

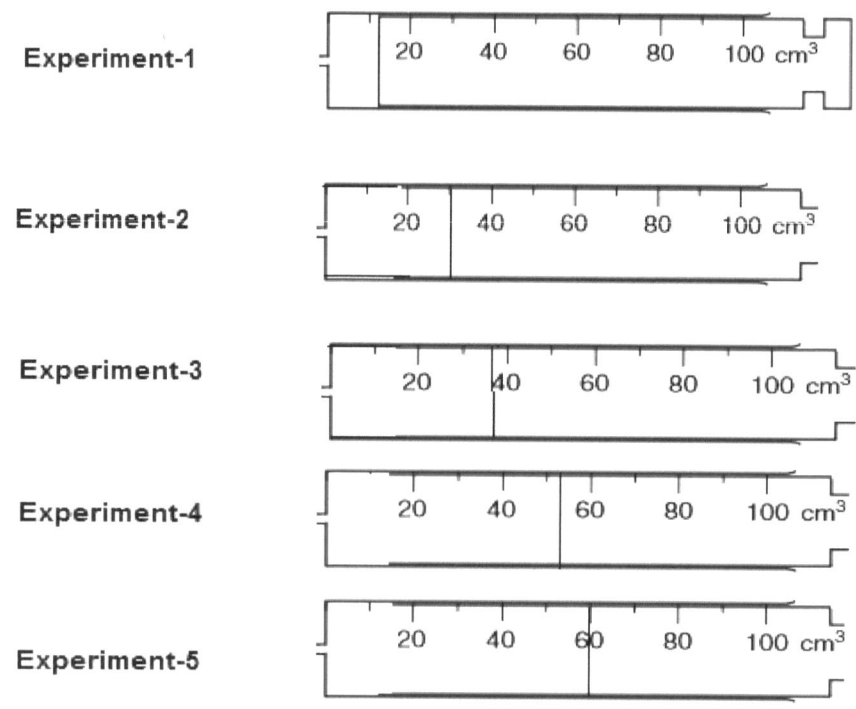

a) Complete the below table of volume of gas collected in cm³ in five experiments

Experiment No.	Volume of gas in cm³
1	
2	
3	
4	
5	

(**b**) Plot the results from experiments 1 to 5 on the grid below and draw a straight line through each set of points.

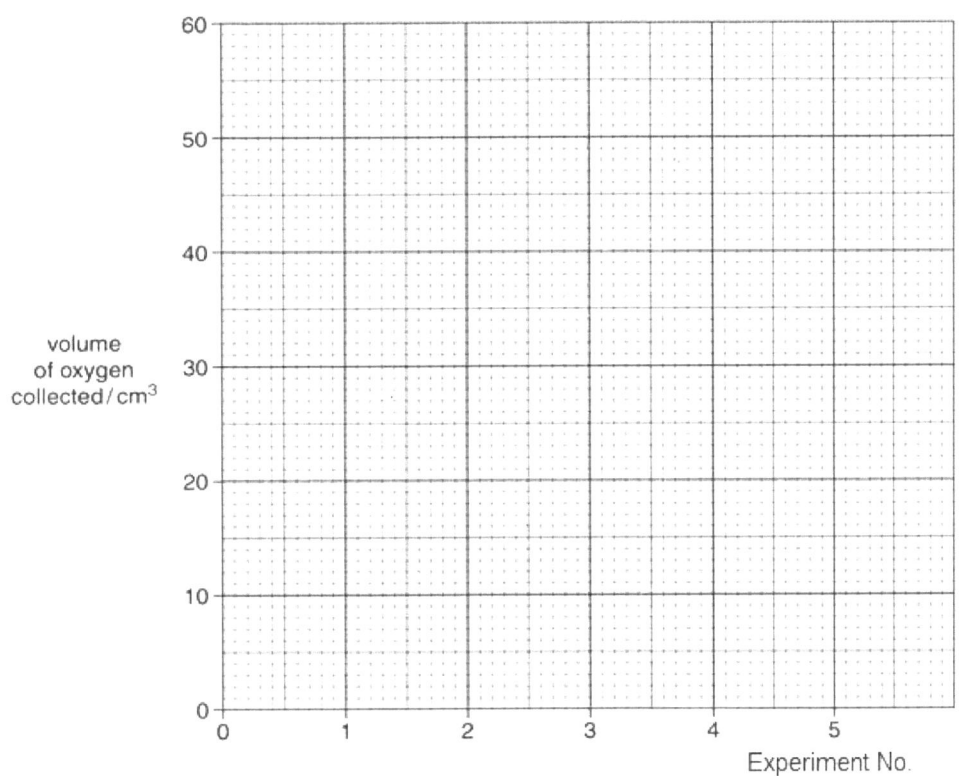

c) Based on your graph explain how volume of oxygen is affected by doubling of reactants each time?

....

....

....

d) Which point on the graph seems to be incorrect and why.

....

....

CHAPTER-3

STRUCTURE OF ATOM

Democritus coined the term 'atom' though not much was known about this elemental particle that made all the matter. The first scientist to explain the composition of matter on the basis of atoms was John Dalton. He proposed the famous Atomic Theory-

Democritus (Alchemist) 400BC

Picture shows John Dalton who first proposed atomic symbols.

According To Daltons Atomic Theory

All matter is made up of very tiny particles called atoms.

Atoms of an element are identical and shape, mass and have similar properties.

Atoms of compounds are different in shape, mass but are always present in a fixed ratio.

During chemical reactions atoms are not formed nor they are destroyed. They only get rearranged to form new products.

Elements: These are the pure substances consisting of same type of atoms. For example Iron, carbon (all elements in periodic table).

Structure of Atom (*Discovery of Sub atomic particles*):

However, the actual study and research about structure of atom began with discovery of sub atomic particles. These particles were discovered by several experiments conducted atomic scientists right from William

Herschel's discovery of heat rays till neutron discovery by James Chadwick. The very first of which was electron.

Electrons: the electrons were first discovered by JJ Thomson by cathode ray experiment. Electrons revolve around the nucleus in spherical orbits. Based on this experiments and findings from the extension of these experiments it was found that electrons have a negative charge (-1) and mass of approx. $1/1847^{th}$ of that of a proton. Therefore, mass of electrons is negligible as compared to the mass of the proton.

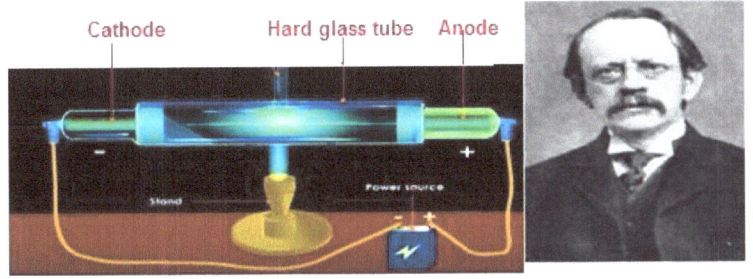

Cathode ray experiment of Thomson   JJ Thomson

Protons: These are the positively charged particles that were first discovered by E Rutherford in 1909 as a part of positively charged nucleus by his gold foil experiment.

Gold Foil Experiment

In this experiment Rutherford scattered alpha particles on a thin gold foil and obtained the results on a fluorescent screen detector. Alpha particles are positively charged Helium atoms (He2+). He observed following.

Observations:

- **Majority of alpha particles went un-deviated.**
- **Few alpha particles got scattered from their original path.**
- **Very few alpha particles rebounded back.**

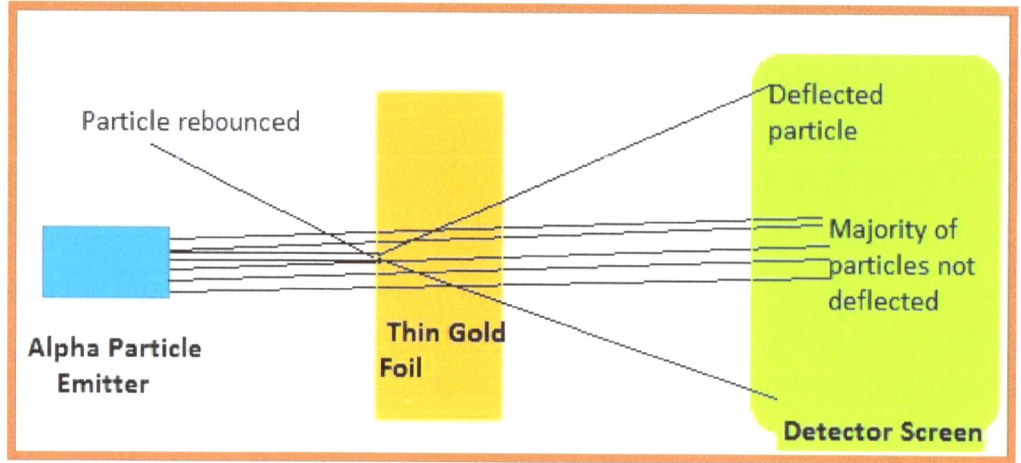

Rutherford's Gold Foil Experiment

Conclusions: Based on above observations Rutherford concluded that:

- Majority of the space of an atom is empty.
- Majority of the mass of an atom and all the positive charge are confined in a small center area, called nucleus. These positively charged particles are protons.
- The total positive charge in the nucleus is equal to the negative charge outside in the form of electrons which revolve round the nucleus in continuous orbits

Neutrons: these are neutral particles that are found in the nucleus of the atoms along with protons. James Chadwick first discovered neutrons. These particles are same in mass as protons (1amu) but have no charge.

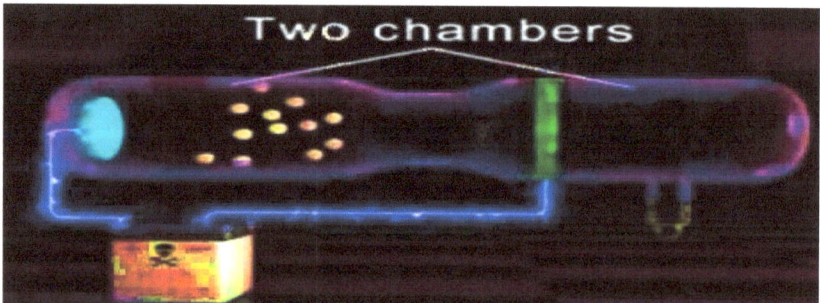

Discovery of Protons Experiment

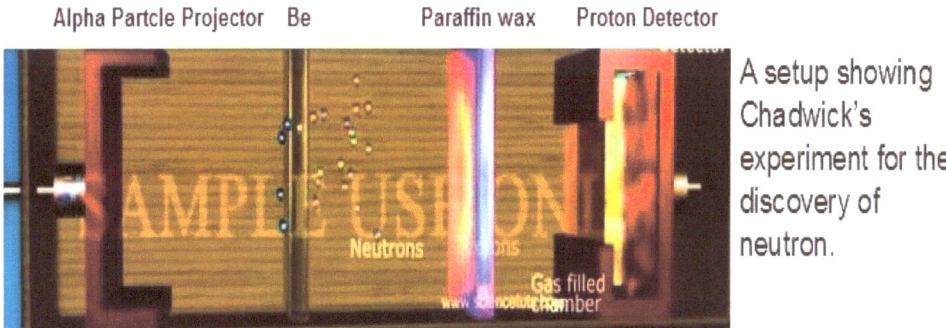

A setup showing Chadwick's experiment for the discovery of neutron.

Following table summarizes the mass and charge of subatomic particles.

Sub atomic Particle	Charge	Mass
Electron	-1	1/1847
Proton	+1	1 (amu)
Neutron	0	1 (amu)

Isotope: These are the atoms of the same element that have different numbers of neutrons e.g. Carbon 12,13 and 14.

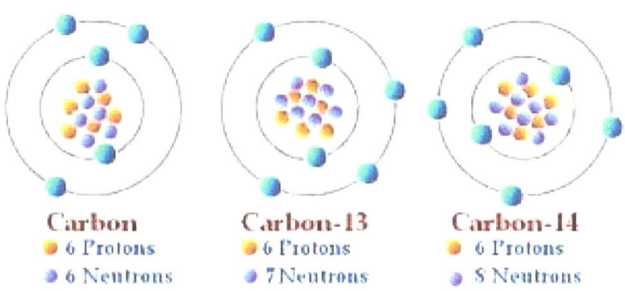

Radio Isotopes: There are non-radioactive isotopes and radio(active)-isotopes. Radioisotopes are unstable atoms, which decay (break down) by giving alpha, beta or gamma radiation. **Medical use**: cancer treatment (radiotherapy) – rays kill cancer cells using cobalt-60.

Industrial use: to check for leaks – radioisotopes called tracers (Americium) are added to oil or gas. At the leaks radiation is detected using a Geiger counter. Another example of radioisotope is carbon 14 – used for carbon dating. When a living tissue dies it does not take in new carbon atoms, but it still has remaining carbon-14 atoms, the radiation can be measured to estimate how long ago fossil died. Radioisotopes were used for producing high yielding crop seeds to increase the agricultural yield.

Electronic configuration / Electronic Arrangement:

Electrons are arranged in electron shells. Atoms want to have full outer shells (full set of valency electrons), this is why they react. Noble gases have full outer shells so they have no need to react. Electron shell structure: 2, 8, 8

Bohr Rules for writing electronic arrangement:

1. Maximum number of electrons that can be accommodated in any shell is given by $2n^2$. Where n is the number of shell. Hence for first shell n=1, therefore only 2 electrons can be placed in the first shell.

2. The maximum number of electrons in last shell of an atom cannot exceed 8.

3. The maximum number of electrons in the second last shell cannot exceed 18.

Using above rules electronic arrangements of elements can be written. For eg. Ca (20) has arrangement 2,8,8,2.

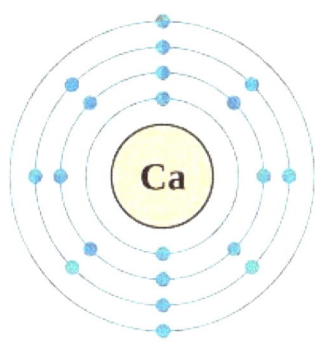

Using above rules if 20 electrons are to be filled in Calcium atom. We can see in the above diagram that 1st shell can have 2. The second and third shell can have 8 each. While the fourth shell has 2 giving electronic configuration of 2,8,8,2.

Technical Terms:

1. Electrons, Protons and Neutrons

2. Atomic number

3. Atomic mass

4. Isotopes

5. Radio isotopes

6. Medical Uses

7. Industrial Uses

Exam Style Questions

Section A –MCQ's

1. Which numbers are added to give the mass number of a cation?

A number of neutrons + number of electrons

B number of electrons + number of neutrons

C number of electrons + number of protons + number of neutrons

D number of protons + number of neutrons

2. Which change to an atom occurs when it forms a mono negative (-1) ion?

A It gains an neutron.

B It gains a electrons.

C It loses an electron.

D It loses a proton.

3. An atom has the symbol $_M X^Y$. Which value determines the position of the element in the Periodic Table?

A M

B Y

C Y+M

D Y - M

4. In the diagrams below, different circles represent atoms of different elements.

Which diagram can represent water vapor?

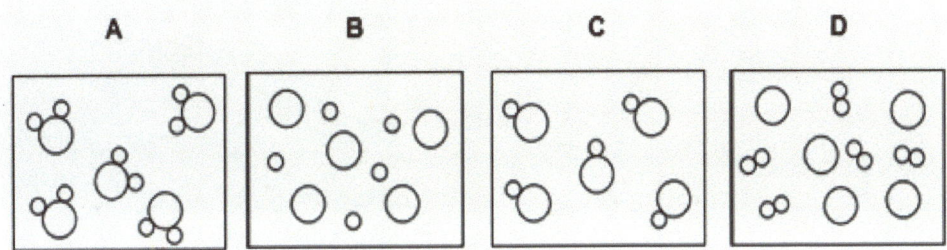

5. Which numbers are added together to give the electron number of an cation?

A number of protons + number of positive charges.

B number of electrons + number of protons

C number of electrons + number of neutrons

D number of protons - number of positive charges

6. The electronic configuration of an ion is 2,8,8. What could this ion be?

A K^+

B Na^+

C Cl^-

D Ca^{2+}

7. Which of the following shows the change that takes place when element Y loses or gains the new particle shown?

		change
A	Electron Gained	an isotope of element Y is formed
B	Electron Lost	the element one place to the right of Y in the Periodic Table is formed
C	Proton Gained	an isotope of element Y is formed
D	Proton Lost	the element one place to the right of Y in the Periodic Table is formed

8. The diagram shows structure of an atom.

What is the proton number and neutron number of the atom?

	neutron number	proton number
A	4	9
B	4	5
C	5	4
D	5	9

9. The symbols of two atoms may be written as shown.

$$^{52}_{24}X \qquad ^{53}_{24}Y$$

Which statement about these atoms is correct?

A They are different elements because they have different mass numbers.

B They are different elements because they have different numbers of protons.

C They are isotopes of the same element because they have the same nucleon number.

D They are isotopes of the same element because they have the same proton number.

9. Element X has seven electrons in its outer shell. How could this element react.

A by gaining two electrons to form a positive ion.

B by losing seven electrons to form a negative ion

C by sharing one electrons with another element to form two covalent bonds

D by sharing two electrons with two electrons from another element to form four covalent bonds

10. What is the relative molecular mass M_r of H_2SO_4.

A 52

B 50

C 98

D 63

Section B – Structured Questions

1. This table shows atoms of some common elements. Complete the table

Element	Proton number	Mass number	Number of		
			proton	electron	neutrons
$_{17}Cl^{37}$					
$_{14}Si^{30}$					
$_{16}S^{32}$					

2. Name and write the symbol of the elements which has the following numbers of particles:

a. 25 electrons, 27 neutrons, 25 protons _____

b. 53 protons, 74 neutrons _____

c. 2 electrons, 2 neutrons (neutral atoms) _____

d. 19 protons, 20 neutrons, 18 electrons_____

e. 86 electrons, 125 neutrons, 82 protons _____

f. 0 neutrons _____

3. Draw a diagram to show the arrangement of the electrons in atom of Neon and Sulphur.

4. a) Complete the below table to show their electron distributions.

Atom	A (atomic number)	Electronic Configuration			
		K shell (#1)	L shell (#2)	M shell (#3)	N shell (#4)
Ca	20	2	8	8	2
K	19				
O	8				
P	15				

5. Below diagram shows some atomic structures without electrons. Complete the structures by filling electrons if X is **Cl⁻** ion and Y is **Ca** atom.

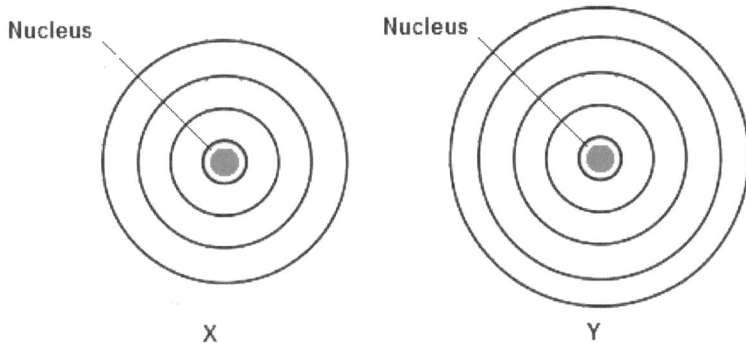

5. Atoms can be identified by atomic numbers and relative atomic mass.

a) What is relative atomic mass.

....

....

b) How is neutron number related to atomic mass.

....

c) What are isotopes.

....

....

d) How are they different from radio isotopes.

....

....

e) Write two medical uses of radio isotopes

....

6. Below diagram shows mass spectrogram of sample of Krypton.

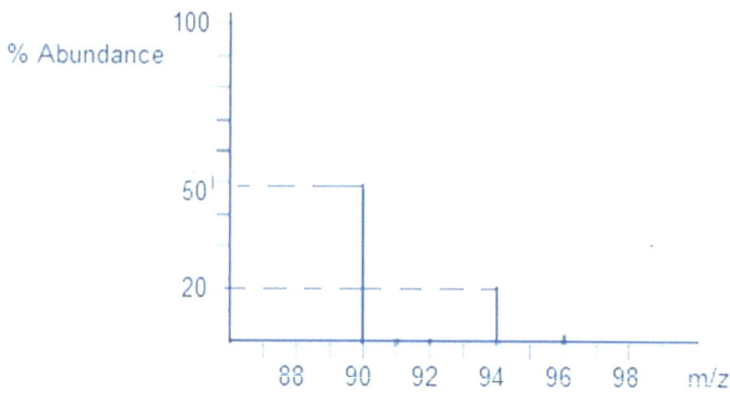

Calculate the average isotopic mass of this sample. Show your working

....

....

....

7. Below are the structures of 4 atoms.

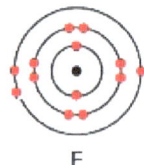 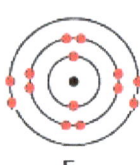 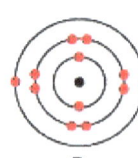 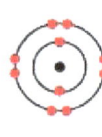

   E   F   G   H

a) Which of these is a Silicon atom and why?

....

b) Which of these is a Neon atom and why?

....

CHAPTER-4

PERIODIC TABLE

History of the Periodic Table

Doberniers Law of Triads : One of the early attempts to classify elements was done by Dobernier in 1817.

According to this law when elements are arranged in increasing order of their masses in the form of triads (three's) such that the mass of center element is the average of the masses of first and third elements, than they have similar properties.

Example: One of the Dobernier's triads.

Element	Symbol	Atomic mass
Lithium	Li	7
Sodium	Na	23
Potassium	K	39

Other scientist tried classifying elements such as Lothermeyer and Newlands but the most remarkable contribution was made by Mendeleev.

Mendeleev first arranged the elements in a table according to their chemical properties in the order of their increasing atomic masses.

His research was further extended by Moseley, who proposed the idea of arranging elements in order of their atomic numbers

Newlands (1865)

No.	No.	No.	No.	No.	No.	No.	No.
H 1	F 8	Cl 15	Co & Ni 22	Br 29	Pd 36	I 42	Pt & Ir 50
Li 2	Na 9	K 16	Cu 23	Rb 30	Ag 37	Cs 44	Os 51
G 3	Mg 10	Ca 17	Zn 24	Sr 31	Cd 38	Ba & V 45	Hg 52
Bo 4	Al 11	Cr 19	Y 25	Ce & La 33	U 40	Ta 46	Tl 53
C 5	Si 12	Ti 18	In 26	Zr 32	Sn 39	W 47	Pb 54
N 6	P 13	Mn 20	As 27	Di & Mo 34	Sb 41	Nb 48	Bi 55
O 7	S 14	Fe 21	Se 28	Ro & Ru 35	Te 43	Au 49	Th 56

Nowadays the elements in the Modern Periodic Table are put in order of increasing atomic number and arranged according to electronic structure.

The modern periodic table is a table containing all elements arranged in ascending order of their atomic number.

Groups:

There are there are 8 groups (vertical Columns) in the periodic table. The group number is equal to the number of electrons in the outer most energy shell of the atoms of the elements in the group. This is the reason why elements of the same group share the similar chemical properties.

Periods:

There are seven periods (horizontal rows) in the periodic table. The period number is also the number of occupied energy shells in the atoms of the elements in the period.

So if an element has 3 valence electrons, it will be in group 3. And if it has 4 energy shells in its configuration, it will be in 4th period.

Metals and Non-metals:

We have two types of elements in the periodic table. These are Metals and Non-metals. As we move in the periodic table from the left to the right, the metallic properties of elements decrease. Metals include Magnesium, Calcium, and Sodium. Non-metals include Carbon, Oxygen and Chlorine. All metals are solid. All non-metals are either solid or gas, except for bromine which is liquid at room temperature.

Metals	Non metals
Metals are generally solid at room temperature except gallium and mercury	Nonmetals are found in all three states of matter
Metals are hard and strong except alkali metals	Nonmetals are brittle and hard.
They generally have high melting and boiling point except mercury	They generally have low melting and boiling point except silica, graphite etc.
They are malleable and ductile.	They are neither malleable nor ductile.
They are sonorous and shiny in appearance.	They are neither sonorous nor shiny in appearance.

Non-Metals

-They have either 4 to 8 valence electrons. Except helium which has two.

-They gain electrons forming negative ions. This tendency is called electronegativity

-They are oxidizing agents as they form positive ions

-They form generally form acidic oxides (CO_2, SO_2) but sometimes neutral oxides (CO and NO).

-Form either ionic compounds with metals, or covalent compounds with other non-metals

Alkali Metals: These are the elements of group 1 of the modern periodic table. They are Lithium, Sodium, Potassium, Rubidium, Cesium and Francium (radioactive). First three of them are; Lithium, Sodium and Potassium. They are called alkali metals as they form strong alkalis on addition to water.

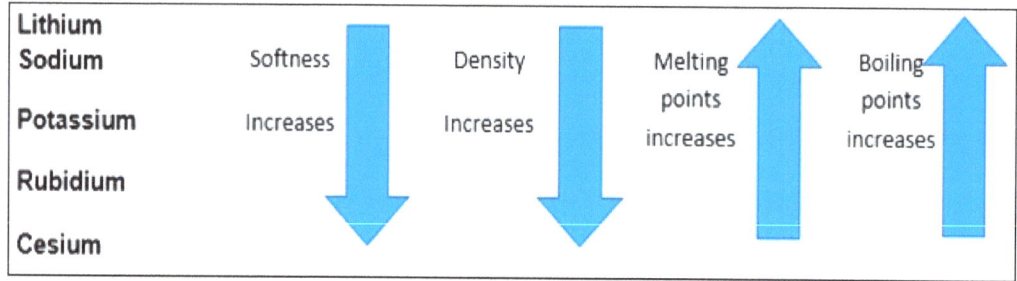

Physical Properties of Alkali metals

They are however, soft. Lithium is the hardest of them and potassium is the softest. They are extremely reactive; they have to be stored away from any air or water. They have low densities and melting points. Like any metals they are all good conductors of heat and electricity.

Chemical Properties of Alkali metals

They react with oxygen or air forming a metal oxide:

$4Na + O_2 \rightarrow 2Na_2O$

Therefore, they are stored under kerosene as they tarnish and get converted to oxide when kept in open air. Their oxides can dissolve in water forming an alkaline solution of the metal hydroxide.

$Na_2O + H_2O \rightarrow 2NaOH$

(Sodium Oxide)   (Water)   (Sodium Hydroxide)

They react with water vigorously forming metal hydroxide and hydrogen gas:

2K + 2H$_2$O → 2KOH + H$_2$

(Potassium)   (Water)   (Potassium Hydroxide)   (Hydrogen)

They React with Halogens forming a metal halide:

2Na + Cl$_2$ → 2NaCl

(Sodium)   (Chlorine)   (Sodium Chloride)

For this group, the further you go down the more reactive the metals become, this is the most reactive group

Alkaline Earth Metals:

These are the elements of group 2 of the modern periodic table. They are Beryllium, Magnesium, Calcium , Strontium, Barium and Radium . They are called alkaline earth metals as they are obtained from alkaline minerals from the earth crust such as calcium is obtained from limestone.

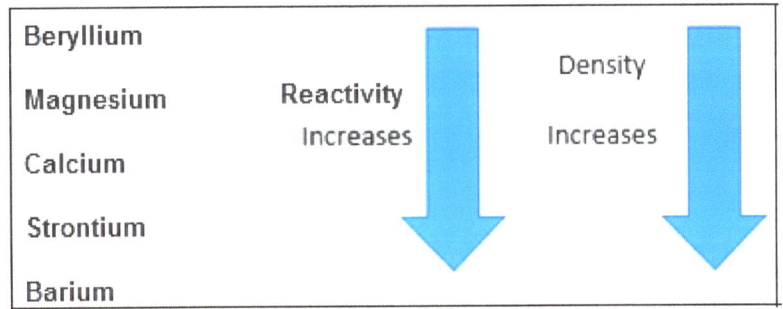

Chemical Properties

They react with oxygen or air forming basic metal oxide:

2Mg + O$_2$ → 2MgO

They react with water forming a basic metal oxide:

Mg + H$_2$O → 2MgO + H$_2$

(Magnesium)   (Water)   (Magnesium Oxide)   Hydrogen

The Halogens

These are elements of group VII and include Fluorine, Chlorine, Bromine, Iodine and Astatine. They are called halogens as they are obtained from sea (Halo = Saline , Gen = to form)

The properties very common halogens chlorine, bromine & iodine. They are colored and the color gets darker as we go down the group. They exist as diatomic molecules (Cl_2, Br_2, I_2). As you go down, they gradually change from gas to solid (chlorine is gas, bromine is liquid and iodine is solid).

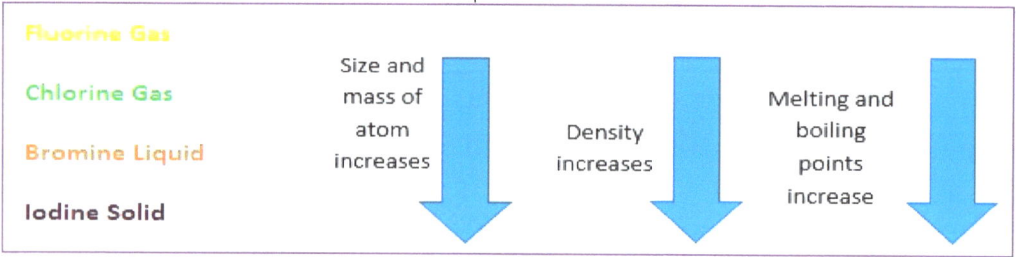

They react with hydrogen forming hydrogen halide, which is an acid if dissolved in water:

H_2 + $Br_2 \rightarrow$ 2HBr

(Hydrogen)   (Bromine)   (Hydrobromic Acid)

They react with metals forming metal halide:

2Zn +   3Cl$_2$ $\rightarrow$ 2ZnCl$_2$

(Iron)   (Chlorine)   (Iron Chloride)

The reactivity also decreases as we do down, Fluorine is most reactive, followed by chlorine bromine then iodine.

If you bubble chlorine gas through a solution of potassium bromide, chlorine will take bromine's place because it more reactive. **This is a displacement reaction.**

2KBr + Cl$_2$ → 2KCl + Br$_2$

(Potassium Bromide) (Chlorine) (Potassium Chloride) (Bromine)

Transition Elements

These are metals present in the center of periodic table from group 3 to 12. Commonly used transition metals are copper, zinc and iron. They have the following properties:

- They are harder and stronger than metals of groups 1 & 2.
- They have much higher densities than other s-block metals.
- They have high melting points except for mercury.
- They are less reactive than metals of group 1 & 2.
- Excellent conductors of heat and electricity.
- They show catalytic activity (act as catalysts)
- They react slowly with oxygen and water
- They show variable oxidation states i.e. They form simple ions with several oxidation states and complicated ions with high oxidation states.
- They also form complex compounds

Colored solutions of Nickel salts. Nickel is a transition metal.

Noble Gases

These are elements of group 8 of modern periodic table. They are colorless gases. They are extremely unreactive; this is because they have their outer energy shell full with electrons that is they have a complete octet of electrons. Therefore, they exist as monoatomic gases in air. So they are stable, this is why they exist as single atoms. They have some uses however, for example argon is used in light bulbs to prevent the tungsten filament from reacting with air, making the bulb last longer. Neon is also used in the advertising and laser beams. These gases are Helium, Neon, Argon, Krypton, Xenon and Radon (radioactive).

Technical Terms

1. Group

2. Period

3. Periodic Trend

4. Alkali metals

5. Alkaline Earth metals

6. Transition metals

7. Halogens

8. Noble gases

Exam Style Questions

1. The letters X, Y and Z represent different atoms.

$$^{52}_{24}Y \quad ^{51}_{25}Z \quad ^{52}_{23}X$$

What can be deduced from the proton numbers and nucleon numbers of X, Y and Z?

A X and Y and Z are the same element.

B X and Z are the same element.

C X has more protons than Y.

D Z has more protons than Y.

2. Which of the following contains the same number of electrons as an atom of neon

A Cl^-

B Li

C Li^+

D N^{3-}

3. Which statement correctly depicts the number of protons in Fluorine and Neon.

A Fluorine has one more than Neon

B Fluorine has one more than Argon

C Fluorine has one less than Neon

D Both have same number of protons.

4. An element Z reacts with hydrogen to from a compound ZH_2. In which of the following groups in the Periodic Table is Z most likely to be found?

A group I

B group II

C group VII

D group VIII

5. Below statements are about four elements M, N, O and Q. Which element is about a transition metal?

A M forms white crystalline chloride.

B N exists in the form of yellow crystals .

C O burns in oxygen to produce heat and white solid.

D Q forms a blue chloride.

6. The O^{2-} ion is different from the O atom. The O^{2-} ion has a

A smaller radius and fewer electrons

B smaller radius and more electrons

C larger radius and fewer electrons

D larger radius and more electrons.

7. Which of these statements correctly describes periodic trends

A Going down the group II reactivity decreases

B Going down the group II atomic size decreases

C Going down the group I atomic size decreases

D Going down the group I reactivity increases.

Section - B: Structured Questions

1. Symbol of Calcium in Periodic Table is $_{20}Ca^{40}$

a) Explain what is meant by the numbers in the above symbol

....

b) Use the numbers to state the number of protons, neutrons and electrons found in Calcium.

....

....

c) Complete the table of the numbers of protons, neutrons and electrons found in the ions shown using your copy of periodic table for atomic numbers and masses

Ion	Protons	Neutrons	Electrons
Li^+			
Mg^{2+}			
Al^{3+}			
Br^-			

d) An isotope of calcium has mass 41. What do you understand by the term isotope?

....

....

2. Rubidium is a metal in Group I of the Periodic Table.

(a) (i) State the number of electrons in the outer shell of a Rb atom.

(ii) How many energy shells are present in an atom of Rb ?

....

(b) An isotope of Rb has a mass number of 88.

(i) What do you understand by the term mass number?

....

(ii) Calculate the number of neutrons in this isotope of Rb.

....

3. Figure below shows part of the Periodic Table of the elements. Use information from Figure to answer the questions that follow. The alphabets are **not** the actual chemical symbols.

(a) Give the symbol for **(i)** an alkaline earth metal,

....

(ii) a halogen and a noble gas.

....

(b) Fluorine, Bromine and Iodine are in same Group (vii). At room temperature Fluorine is a gas and Bromine is a liquid. Predict whether Iodine at room temperature, is a gas, liquid or a solid.

....

(c) Explain the trend in reactivity in Group VI by the help of a chemical reaction.

....

....

(d) Write the formula for a compound that is formed when

(i) An element Y from Group I reacts with an element Z from Group VI,
....

(ii) An element Y from Group II reacts with an element X from Group VI.
....

4. Neon has atomic number 10, one more than Fluorine. Explain why:

a. Fluorine forms diatomic molecules, while Neon is monatomic

....

....

b. Fluorine is reactive, but Neon is unreactive

....

....

c. Both Fluorine and Neon are gases at room temperature

....

....

5. Strontium lies below calcium in Group II, and next to rubidium in Group I. Explain why it is:

a. More reactive than calcium

b. Less reactive than rubidium

c. Barium lies below strontium in Group II. What can you conclude about its chemical properties?

d. Strontium belongs to the Vth Period of the Periodic Table. Explain how the atomic size of elements change on moving across a period from left to right.

6. Complete the following table to estimate the boiling point of cesium and predict the reactivity of cesium with water.

Metal	Mass in (amu)	Melting point (^{0}C)	Boiling point (^{0}C)	Density (g/cm3)
Li	6.0		1,342	0.53
Na	22.9	96.9	883	0.97
K	39.0	63.1	759	0.89
Rb	85.5	39.8	688	
Cs	133	28.4		1.93

a) Predict the melting point of lithium.

....

b) Predict the density of Rubidium.

....

c) Explain the trend in melting and boiling point in group –I.

....

....

....

d) i- Write a balanced reaction of Rubidium with water.

....

ii) Predict the density of rubidium

....

e) Explain the trend in density in group –I.

....

....

h) Explain by the help of diagrams why valency of carbon is 4 but that of Fluorine is 1.

....

....

....

CHAPTER-5

STOICHIOMETRY /MOLE CALCULATION

Elements: These are the pure chemical substances that have same kind of atoms. For example carbon (C), nitrogen (N), calcium (Ca). Periodic table is arrangement of atoms by their atomic numbers.

Compounds: These are the substances made of atoms of different elements combined in a fixed ratio. For example water (H_2O), carbon dioxide (CO_2) etc.

Chemical Formula (Molecular Formula): It is a formula that represents actual number of atoms of different elements in one molecule of the compound. For example formula of calcium carbonate is $CaCO_3$ (this represents that each unit of calcium carbonate contains one atom of calcium, one atom of carbon and three atoms of oxygen).

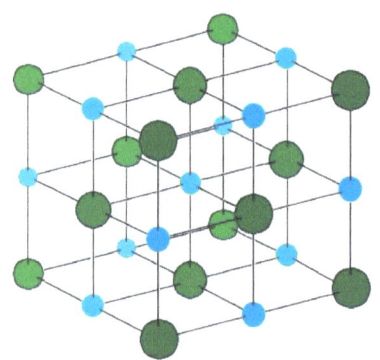

Diagram shows lattice structure of NaCl where blue particles and Cl^- ions and green particles are Na^+ ions

Example-What is the formula of sodium chloride that has above structure.

Solution-The above diagram shows that one unit of sodium chloride contains $Na13Cl13$. Therefore its formula should be NaCl (simplest form).

Construction of Symbol and Word Equations: An equation is a symbolic representation of a chemical reaction. For example when

Magnesium reacts with hydrochloric acid following equations can be written.

Word Equation: Magnesium + Hydrochloric acid —> Magnesium chloride + hydrogen.

Symbol Equation: $Mg(s) + 2HCl\ (aq) \longrightarrow MgCl_2\ (aq) + H_2\ (g)$

What is Stoichiometry..?

The branch of chemistry that deals with the study of measuring the amount of substances (reactants and products) involved in a chemical reaction is called stoichiometry.

Balancing of Equations

• Equations are should always be written in balanced form as per law of conservation

of mass. According to this law "during a chemical reaction the mass of the reactants should always be equal to the mass of the products".

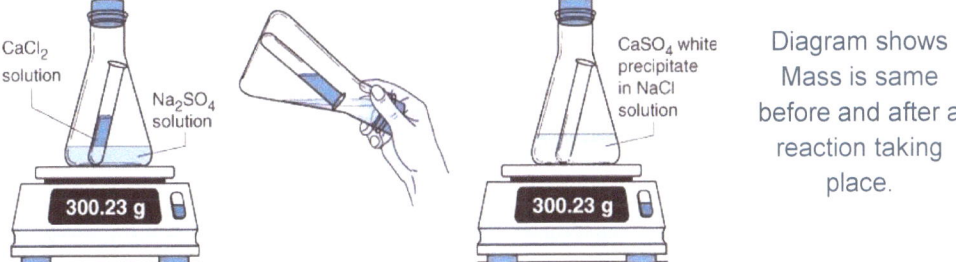

Rules for balancing equations:

1. -In a symbol equation the number of atoms on each side of the equation should

balanced for each element.

2-In an equation, remember to write the symbol for the state (aqueous, solid, liquid, gas) in brackets, and the oxidation state of a transition metal in a word equation.

3. After writing the equation identify metals and nonmetals in the reaction.

4. Numbers are added as coefficients in front of the formula and never in between or after the formula.

5. Than balance the metals on both sides first.

6. Balance the nonmetals except hydrogen and oxygen after all metals are balanced

7. Balance hydrogen (if any) now.

8. Balance oxygen (if any) now.

Let's apply the above rules and balance some reactions now. Balance below reaction

as per rules for balancing

Example- $ZnO + H_3PO_4$ ------------- $Zn_3(PO_4)_2 + H_2O$

Solution-Balance Zinc (metal) first- Add a 3 in front of ZnO

Balance Phosphorus (nonmetal) now- Add a 2 in front of H_3PO_4. Balance hydrogen –Add a 3 in front of H2O now. Finally we see that oxygen gets balanced automatically (following the rules sequence)

$3ZnO + 2H_3PO_4 \longrightarrow Zn_3(PO_4)_2 + 3H_2O$.

Now try and balance some equations yourself-

Eg-2 ____ $Pb(OH)_2$ + ____ HCl ____ H_2O + ____ $PbCl_2$

Eg-3 ____ Na_3PO_4 + ____ $CaCl_2$ ____ NaCl + ____ $Ca_3(PO_4)_2$

Relative Atomic mass (RAM) : It is defined as the mass of an atom any element as compared to the mass of 1/12th of a carbon -12 atom. The masses of elements in periodic table are their relative atomic masses as shown in the below excerpt of IGCSE periodic table.

I	II					
6.9 Li lithium 3	9.0 Be beryllium 4					
23.0 Na sodium 11	24.3 Mg magnesium 12					
39.1 K potassium 19	40.1 Ca calcium 20	45.0 Sc scandium 21	47.9 Ti titanium 22	50.9 V vanadium 23	52.0 Cr chromium 24	54.9 Mn manganese 25
85.5 Rb rubidium 37	87.6 Sr strontium 38	88.9 Y yttrium 39	91.2 Zr zirconium 40	92.9 Nb niobium 41	95.9 Mo molybdenum 42	– Tc technetium 43
133 Cs caesium 55	137 Ba barium 56	139 La lanthanum 57	178 Hf hafnium 72	181 Ta tantalum 73	184 W tungsten 74	186 Re rhenium 75

Key: relative atomic mass / atomic symbol / name / atomic number

Naming Compounds:

1- For positive part / ion the name remains same. For e.g. sodium oxide (Na_2O)

2- For negative part (ion) the name is suffixed. For e.g. oxide in Na_2O

For covalent bonds, prefixes are used to denote the number of atoms.

1 = mono (carbon monoxide)   2 = di (carbon dioxide)

3 = tri (phosphorus trihydride)   4 = tetra

5 = penta   6 = hexa

3- The only time we drop a prefix is if the mono is to appear at the beginning of the name. If there is an oxide the 'a' or 'o' of the prefix is lost e.g. carbon monoxide.

4-If a metal ion is combined with a polyatomic ion in a compound and one is oxygen, the name ends in –ate, except hydroxides

5-With ionic compounds, the cation (metal) goes first in the name. With covalent compounds the element further on the left goes first (hydrogen is thought of being in between nitrogen and oxygen so: sodium chloride phosphorus trihydride/hydrogen peroxide)

Information conveyed from a chemical equation:

A balanced chemical equation conveys a lot of information. For example for the balanced reaction as shown

$2\ AlBr_3 + 3\ K_2SO_4 \longrightarrow 6\ KBr + 1\ Al_2(SO_4)_3$

The information conveyed is-

- There are two reactants and two products in this reaction.

- 2 moles of Aluminum bromide react completely with 3 moles of potassium sulphate.

- Total 6 moles of potassium bromide and 1 mole of aluminum sulphate are formed in this reaction.

Definition of Mole: One mole is the amount of substance that contains 6.02×10^{23} atoms or molecules of that substance.

The Mole Concept

Three statements can explain this concept.

1. One mole of an element = gram atomic mass of that element = 6.02×10^{23} atoms

For eg. 1 mole of carbon = 12gm (from periodic table) = 6.02×10^{23} atoms of carbon.

2. One mole of a molecule = gram molar mass of that of that compound = 6.02×10^{23} molecules of that compound.

3. One mole of a gas = gram atomic /molar mass of that gas = 6.02 X 10^{23} particles of that gas = 24 dm3 at RTP (room temperature and pressure) This 24dm3 is known as molar volume.

Molar Mass- It is the mass of a molecule of a substance as compared to the mass of 1/12th of carbon -12 atom.

To find the molar mass (Mr) of a substance, we add together the relative atomic mass for all the atoms shown in its chemical formula times the number of each atom.

Example 1

What is the relative formula mass of hydrogen sulphide, H2S?

(Ar of H = 1, Ar of S =32)

Mr of H2O = 1 + 1 + 32 = 34

Example 2

What is the relative formula mass of magnesium hydroxide, Mg(OH)2?

(Ar of Mg = 24, Ar of O = 16, Ar of H = 1)

Mr of Mg(OH)2 = 24 + (16x2) + (1x2) = 58

Important Formulae

1. Moles = Mass / Molar mass

2. Moles = No. of particles / 6.023 x 10^{23}

3. Moles = Volume (at RTP) / 24 dm^3 (only applies for gases)

Example

a) What mass of carbon dioxide is made from 84 g of magnesium carbonate? b) Also calculate the mass of Magnesium oxide that can be formed from 8.4g of magnesium carbonate.

Solution: Part a)-Write an equation for the said change

$MgCO_3 \longrightarrow MgO + CO_2$

Now assign molar masses and compare

$MgCO_3 \longrightarrow MgO + CO_2$
  84g   40g   44g

Thus 84g of carbonate will give 44g of CO_2 = 2 x 22 = 44 g

$$MgCO_3 \longrightarrow MgO + CO_2$$

Part b) From balanced reaction   84   40   44

Therefore from 8.4 g of carbonate will be give 4g of Magnesium oxide.

Limiting Reagent: In a given reaction the reagent which is comparatively less in amount is called limiting reagent and the other reagent which is comparatively more in amount is called excess reagent.

Remember- Limiting reagent decides the amount of products always.

Reacting Masses And Ratios

You should be able to calculate the mass of a product or reactant using the idea of moles, a balanced equation, limiting reagent and relevant Ar values.

Example: Hydrochloric acid and sodium hydroxide react together to make sodium chloride and water:

$HCl + NaOH \rightarrow NaCl + H2O$

What mass of sodium chloride is made when 20 g of sodium hydroxide reacts with excess acid?

Solution – Here question itself says acid is taken in excess thus sodium hydroxide is the limiting reagent.

From balanced equation with molar masses

HCl + NaOH → NaCl + H2O
36.5   40   58.5   18

Now by comparing the masses since 40g of hydroxide gives 58.5 g of NaCl, 20 g of hydroxide will give

$$\frac{58}{2} = 29.25 \text{ g}$$

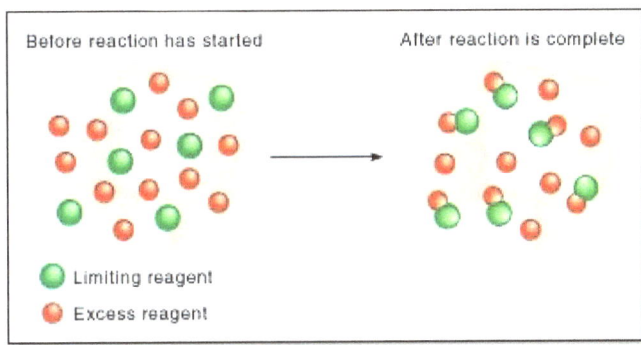

Diagram shows illustration-1 to show how reactants prove to be limiting or excess during chemical changes.

Question 1
How many burgers can be made?

Which part is the limiting reagent? (the part with the smallest amount)

Diagram shows illustration-2 to show how reactants prove to be limiting or excess during chemical changes. Can you tell why only three burgers be made ?

*Molar Volume :*Volume of one mole of **any gas is 24dm³ at RTP** (RTP = room temperature and pressure; 25°C and 1 atm)

Calculations involving Molar volume of gases

*Example 1:*Calculate the volume of 0.5 mol of nitrogen dioxide at rtp.

Solution- volume = 0.5 × 24 = 12 dm3

For volume conversions remember that -

1dm3 =1 Liter =1000 cm³ =1000 ml

*Example 2 :*Calculate the volume of CO_2 gas produced when 20cm³ of methane burns in 20cm³ of oxygen gas.

Solution- Write a balanced reaction for the above first

CH4(g) + 2O2(g) → CO2(g) + 2H2O(l)

Remember- volumes can be directly compared like moles.

Therefore 20cm³ of methane as per balanced reaction needs 40cm³ of oxygen thus here oxygen is the limiting reagent.

Also since 2 moles of oxygen produces only 1 mole of carbon dioxide 20cm³ of oxygen will produce only 10cm³ of carbon dioxide. Hence the answer is 10 cm³

Concentration of Solutions

Many chemicals react in an aqueous solution form, therefore direct masses of these substances can't be used for calculation. Hence their masses in solutions are considered in the form of Molar concentration.

Molar concentration is defined as the number of moles of solute dissolved in 1 dm3 of solution.

Important Formulae-

1. Molar Concentration = No of moles / Volume in dm3

2. Molar Concentration = $\dfrac{\text{Mass}}{\text{Molar mass}} \times \dfrac{1000}{\text{Volume in cm3}}$

Example:

When 20g of magnesium is treated with 100cm³ of 0.5 mol/dm³ sulphuric acid.

a) Determine the limiting reagent?

b) Determine the mass of magnesium sulphate that can be produced?

Solution a) first we write the balanced reaction.

Mg + H$_2$SO$_4$ ———> MgSO$_4$ + H$_2$

20g + 100cm³/0.5M

Compare the moles of reactants

Mg	+	H$_2$SO$_4$
20/24		0.5 x0.1
0.83		0.05

a) Thus acid is the limiting reagent.

Hence it will decide the products. Therefore 0.05 moles of magnesium sulphate would be obtained.

b) 0.05 x 120 (molar mass of MgSO$_4$) =6g

Water of Crystallization /Hydration

The water molecules present in hydrated salts such as copper sulphate pentahydrate CuSO4.5H2O are called water of hydration. The ratio of moles can be compared to find the amount of water of hydration in compounds as shown by following example.

Example

24.6 grams of a hydrated salt of MgSO$_4$.xH$_2$O, gives 12.0 g of anhydrous

$MgSO_4$ on heating. What is the value of x?

First we find the mass of water driven off.

Mass of water evolved = 24.6 – 12.0 = 12.6 g

Therefore moles of water involved 12.6/18 = 0.7

Also moles of Magnesium sulphate involved 12/120 = 0.1

Since ratio is 1:7 we get formula of hydrated salt as $MgSO_4.7H_2O$

Experiment below shows dehydration of copper sulphate penta-hydrate (blue crystals) changing colour to white on heating. The process can be reversed by adding water again to anhydrous copper sulphate.

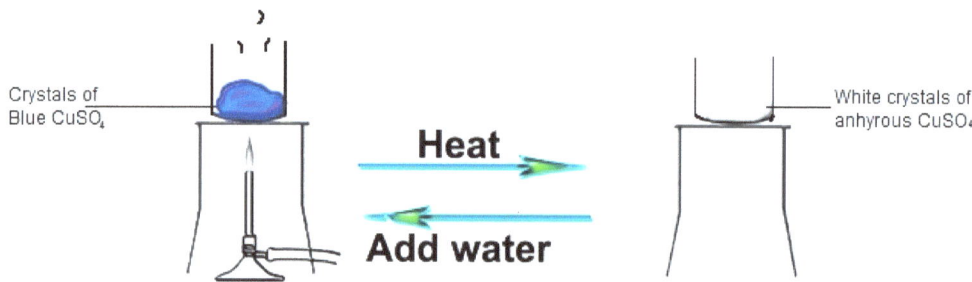

Empirical and Molecular Formula

Empirical Formula : It is the formula that represents a simple whole number ratio of atoms of different elements present in one molecule of a compound. For example the empirical formula of benzene (C6H6) is CH.

Molecular Formula: It is the formula that represents the actual number of atoms of different elements of present in one molecule of that compound. For example the molecular formula of benzene is C6H6.

Example -A compound A contains 73.47% carbon and 10.20% hydrogen by mass, the remainder being oxygen. It is found from other sources that A has a Relative Molecular Mass of 98 g mol-1. Calculate the molecular formula of A.

Solution- It is not necessary to put in all the details when you carry out a calculation of this type. The following is adequate.

Elements	C	H	O
Percentage	73.47%	10.20%	16.33%
Mass in 100g of compound	73.47g	10.20g	16.33g
Mole Ratio	$\frac{73.47}{12}=6.12$	$\frac{10.20}{1}=10.20$	$\frac{16.33}{16}=1.020$
Simple ratio (divide by smallest of all)	$\frac{6.12}{1.02}=6$	$\frac{10.20}{1.20}=10$	$\frac{1.02}{1.02}=1$
Empirical Formula	C= 6	H = 10	O =1

Therefore, the empirical formula is $C_6H_{10}O$. To find molecular formula:

n = $\frac{\text{Empirical Formula Mass}}{\text{Molecular Formula mass}}$

In this case n = 98 /98 = 1

Molecular Formula = Empirical Formula Mass x 1

The molecular formula is the same as the empirical formula $C_6H_{10}O$.

Percentage Yield:

It is defined as the percentage ratio of actual yield to theoretical yield of the product in a chemical reaction. It is calculated by following formula

$$\text{Percentage Yield} = \frac{\text{Actual Yield}}{\text{Theoretical Yield}} \times 100$$

Example: In a reaction of Zinc and nitric acid. If a sample of 40g impure zinc completely reacts with excess of nitric acid. Such that 80g of zinc nitrate salt is produced. Determine the percentage yield of the salt.

Solution: Note that the yield that is obtained is always actual yield. Thus to calculate the percentage yield we need to find the theoretical yield.

From balanced reaction:

$$Zn + 2HNO_3 \longrightarrow Zn(NO_3)_2 + H_2$$

Now as per the question theoretical moles of Zn that reacted are 40/65 = 0.61 moles Therefore as per reaction stoichiometry 0.61 moles of zinc nitrate should be formed. Which is 0.61 x 189g(molar mass of Zinc nitrate), which is equal to 115.29 g

Now Theoretical yield = 115.29g, Actual yield = 80g (normally always given in Question)

Therefore percentage yield = 80 / 115.29 x100 = 69.39%.

Percentage Purity

It is calculated by following formula

$$\text{Percentage Purity} = \frac{\text{Mass of Pure sample}}{\text{Mass of Impure sample}} \times 100$$

Example: An impure sample of 200g hydrated Iron Sulphate (FeSO4.7H2O) when heated produced 14g of water (loss in mass in crucible was weighed). Calculate its percentage purity.

Solution: To calculate the mass of pure sample we find the number of moles of water lost and relative number of moles of pure compound.

We then convert it to mass in gm and find the purity from above formula 14 g of water is 0.77 moles of water. Therefore the moles of actual salt are 0.77 / 7 = 0.11 (as each mole contains 7 water of crystallization)

Thus mass of pure sample is 0.11 x 430 (molar mass of hydrated salt) = 47.3 g

Thus mass of pure salt in the sample was only 47.3g

Therefore percentage purity .

$$= \frac{47.3}{200} \times 100 = \mathbf{23.65\ \%}$$

Conceptual Practice Questions:

1. In the compounds given below, state the name of atoms and their respective ratios.

i- Methane CH_4

....

ii- Ammonia NH_3

....

iii- Silver chloride AgCl

....

iv- Calcium carbonate CaCO₃

....

v- Potassium magnate (VII) KMnO₄

....

2. Balance the following equations by adding the coefficients in blank spaces only.

a. ____ Li₃PO₄ + ____ KOH ——> ____ LiOH + ____ K₃PO₄

b. ____ Sn(OH)₂ + ____ HCl ——> ____ H₂O + ____ SnCl₂

c. ____ As + ____ O₂ ——> ____ As₂O₃

d. ____ B + ____ HCl ——> ____ H₂ + ____ BCl₃

e. ____ HI + ____ Mg(OH)₂ ——> ____ MgI₂ + ____ H₂O

3 a. Explain what is meant by the term *relative atomic mass*?

....

....

b. Work out the relative molecular masses of the following compounds.

(i) Calcium carbonate

....

(ii) Ammonium hydroxide.

....

(iii) Propanol (C_3H_7OH)

....

(iv) Sodium bromide.

....

(v) Magnesium sulphate.

....

....

4. Find the concentration of a LiOH solutions in two bottles A and B that have labels as follows.

a. Label on bottle A -0.4 moles in 5 dm^3 of solution?

....

....

b. Label on bottle B -0.7 moles of in 400 cm^3 of solution?

....

....

5. Find the number of moles of a NaCl solute in following solutions:

a. 1.5 dm^3 of a 0.03 mol/ dm^3 solution.

....

b. 250 cm^3 of a 2.5 mol/ dm^3 solution.

....

6. Calculate the volume of following gases at rtp :

a) 14 g of Sulphur dioxide, SO_2

b) 100 g of nitrous oxide N_2O

7. How much carbon dioxide is produced by complete combustion of 128 g of ethane? ($2C_2H_6 + 7O_2 \longrightarrow 4CO_2 + 6H_2O$)

....

....

Exam Style Questions

Section A-MCQ Questions

1. Below are the two statements about moles.

i- Moles in 24g of carbon and 40g of calcium are same.

ii- 400ml of N_2 gas has same number of particles as 400ml of O_2 gas.

Which statement correctly describes i and ii.

A i is correct ii is incorrect

B i is incorrect ii is correct

C Both i and ii are correct

D Both i and ii are incorrect

2. What is the number of molecules in 240 cm³ of oxygen under room temp conditions?

A 1.25×10^{22}

B 1.34×10^{22}

C 6.0×10^{20}

D 6.0×10^{21}

3. The equation for the reaction of hydrogen in Selenium is
$H_2 + Se \longrightarrow H_2Se$

This equation indicates that

A 2 atoms of hydrogen combine with 2 atoms of Selenium

B 2 moles of product can be obtained form 1 mole of Selenium

C 2 moles of product can be obtained from 2g of hydrogen

D 1 mole of product is obtained from 2g of hydrogen.

4. One mole of nitrogen N_2 has a mass of

A 0.028 kg **B** 28 kg **C** 0.028g **D** 0.28g

5. An instruction from a lab manual is as shown

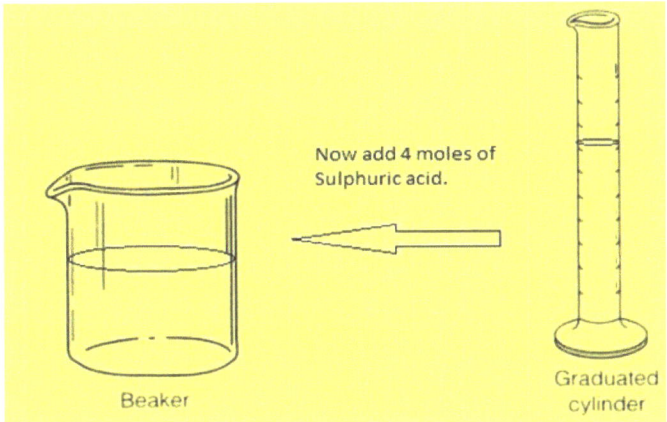

If the molar concentration of acid used is $2mol/dm^3$, how many dm^3 of acid should be added as per the above instruction?

A $1 \, dm^3$

B $2 \, dm^3$

C $3 \, dm^3$

D $4 \, dm^3$

6. The mass of one mole of a chloride formed by a metal 'X' is 58.5 g. The formula of the chloride could be

A XCl_2

B XCl

C X_2Cl

D X_3Cl

7. 200 cm³ of gaseous hydrogen contains y molecules. How many molecules are there in 400 cm³ of gaseous methane under the same conditions of temperature and pressure?

A y

B 2y

C y/2

D 8y/3

8. The relative molecular mass of Bromine is 80. What is the mass of 4 moles of bromine gas (Br_2)?

A 160 g

B 640 g

C 320 g

D 180 g

9. If 2 moles of Nitrogen is oxidized to Nitrogen dioxide, how many moles of oxygen is required? ($N_2 + 2O_2 \rightarrow 2NO_2$)

A 1.0   **B** 2.0   **C** 4.0   **D** 1.5

10. A metal X forms a sulphite of formula $X_2(SO_3)_3$. What will be the formula of its nitrate?

A X_5NO_3

B $X_2(NO_3)_5$

C $X_2(NO_3)_3$

D $X(NO_3)_3$

11. A compound contains 81.71% carbon and 18.29% hydrogen by mass. What can be possible molecular formula of this compound?

A C_2H_6

B CH_4

C C_3H_6

D C_3H_8

12. Barium oxide is produced by heating barium carbonate as shown.

$$BaCO_3 \text{ (s)} \longrightarrow BaO \text{ (s)} + CO_2 \text{ (g)}$$

What mass of Barium oxide will be produced on heating one metric ton of barium carbonate?

A 153/100 metric tons

B 197/100 metric tons

C 197/153 metric tons

D 153/197 metric tons

13. The empirical formula of a liquid is CH. Which other information is needed to work out its molecular formula?

A boiling point

B relative molecular mass

C density

D volume occupied by 1 mole

14. Potassium is an alkali metal found in group –I of periodic table.

Potassium reacts with water as per the given equation

2K + 2H$_2$O -----> 2KOH + H$_2$

Which volume of hydrogen is produced at rtp when 0.8 mol of potassium reacts?

A 1.2 dm^3

B 4.8 dm^3

C 2.4 dm^3

D 9.6 dm^3

Structured Questions:

1. Sulfur combines with Zinc when heated to form the pale crystals of Zinc(II) sulfide:

$$Zn\ (s) + S\ (l) \longrightarrow ZnS\ (s)$$

In the above experiment, 6.74 g of Zn is allowed to react with 9.14 g of Sulphur.

a) Determine the limiting reagent, and what is the reactant in excess?

....

....

....

....

b) Calculate the maximum mass of ZnS that can be formed.

....

....

2. When Copper(II)hydroxide is mixed with sulphuric acid, copper (II) sulphate precipitate results.

$Cu(OH)_2$ (aq) + H_2SO_4 (aq) ---> $CuSO_4$ (s) + H_2O(l)

a) If 3.20 g of $Cu(OH)_2$ is treated with 2.50 g of sulphuric acid, what is the limiting reagent and what is the reactant in excess?

....

....

....

....

b) How many grams of copper (II) sulphate precipitate can be formed?

....

....

c) If 1.45 g of $CuSO_4$ is actually obtained, what is the percent yield?

....

....

d) If 450 cm³ of 0.4M HCl completely neutralizes 200cm³ of KOH solution.

Calculate the molar concentration of this KOH solution.

....

....

....

3. When calcium carbonate is strongly heated above 100°C it decomposes.

$CaCO_3$ (s) ---> CaO (s) + CO_2 (g)

a) What is meant by (s) and (g)?

....

b) Describe a chemical test to identify the carbon dioxide produced.

....

....

c) Calculate the relative molecular mass (Mr) of calcium oxide.

....

....

d) Calculate the mass of calcium oxide formed when 8.60 g of calcium carbonate is heated.

....

....

4 a). What volume of 1.2 mol/dm³ nitrous acid will react with 10 g of calcium carbonate?

$CaCO_3$ (s) + $2HNO_2$ (aq) ---> $Ca(NO_2)_2$ (aq) + H_2O (l) + CO_2 (g)

....

....

b. When the reaction in above experiment is over (at RTP) determine the volume of CO_2 that can be produced.

....

5. The Grignard reagent used in organic synthesis has a formula C_2H_5MgBr.

(a) Calculate the percentage composition by mass of C_2H_5MgBr by the following steps.

(b) i- The number of moles of Mg in one mole of C_2H_5MgBr = ……….

ii- mass of Mg in one mole of compound ………………g

(c) The mass of one mole of C_2H_5MgBr = ……………...g

(d) Percentage composition of Mg in compound =

$\dfrac{\text{Mass of Mg in compound}}{\text{Molar mass of } C_2H_5MgBr} \times 100$

……………………………………………………………..

6. Nickel (II) nitrate contains water of crystallization and the full formula of the salt is of the type $Ni(NO_3)_2 \cdot xH_2O$:

1 mole of $Ni(NO_3)_2 \cdot xH_2O$ will produce 1 mol of Nickel (II) oxide The following experiment was carried out to find "x".

A 5.0 g sample of the salt $Ni(NO_3)_2 \cdot xH_2O$ was heated to leave 1.27 g of Nickel(II) oxide.

(a) What is meant by the phrase one mole of a substance?

……………………………………………………………………………

……………………………………………………………………………

(b) How many moles of NiO there in 1.27g of Nickel (II) oxide?

……………………………………………………………………………

……………………………………………………………………………

(c) How many moles of Ni(NO₃)₂. xH₂O are there in 5.0g of hydrated Nickel (II) nitrate?

....

....

(d) Calculate the mass of one mole of Ni(NO₃)₂. xH₂O.

....

....

(e) The mass of one mole of Ni(NO₃)₂ is 183g. Calculate x.

....

....

7: An organic compound contains carbon 40.00 %, oxygen 53.285 % and hydrogen 6.714 %. What is the molecular formula of the compound if its molar mass is 90g /mol ?

....

....

....

....

8. When a mixture of bicarbonates containing sodium and magnesium bicarbonate was treated with sulphuric acid bubbles of colorless gas carbon dioxide are seen as shown on the next page.

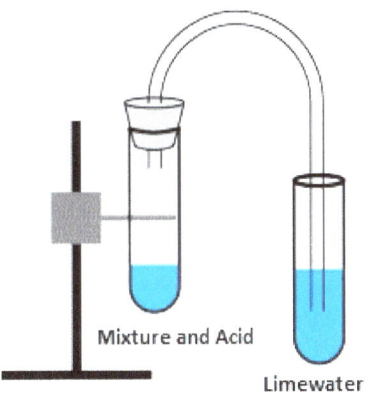

Mixture and Acid

Limewater

$NaHCO_3 + H_2SO_4 \longrightarrow Na_2SO_4 + H_2O + CO_2$

$Mg(HCO_3)_2 + H_2SO_4 \longrightarrow MgSO_4 + H_2O + CO_2.$

a) Rewrite the above equations in a balanced form.

....

....

b) If 0.5 mole/dm3 of 1.2 dm³ of H_2SO_4 completely reacts with the mixture.

i- Calculate the number of moles of acid reacted.

....

....

ii- Based on the reaction stoichiometry calculate the number of moles of each carbonate in the mixture.

....

....

c) Calculate the total number of moles of water produced in the above reaction.

d) Calculate the total number of moles of carbon dioxide produced in the above reaction.

....

....

9. When ethane burns in excess of oxygen 100 cm³ sample of ethane, C_2H_6, is burnt in an excess of oxygen.

$$2C_2H_6(g) + 7O_2(g) \longrightarrow 4CO_2(g) + 6H_2O(l)$$

(a) What is the volume of carbon dioxide is produced? ………………..

(b) What volume of oxygen is required? …………………………………

(c) Calculate the mass of water formed if volumes are measured at RTP.

....

....

(d) Another hydrocarbon gas of the same series (alkanes) was used for combustion in place of ethane such that at constant conditions 200 cm³ of gas burnt completely in 1.3 dm³ of oxygen, producing 800 cm³ of CO_2. Determine the formula of the gas. Show your working

....

....

....

....

Technical Terms:

1. Mole

2. Molar concentration

3. Limiting reagent

4. Percentage composition

5. Percentage yield

6. Percentage purity

7. Empirical Formula

8. Molecular Formula

9. Molar volume

10. Molar ratios

11. Molar concentrations

CHAPTER- 6

CHEMICAL BONDING

Chemical Bond

It is a strong force of attraction between two or more atoms that results in the formation of a molecule or compound. Elements bond with each other to complete their octet (eight electrons in the outermost shell). This was suggested by Kossel and Lewis. Therefore it is also known Kossel and Lewis "Octet Rule".

  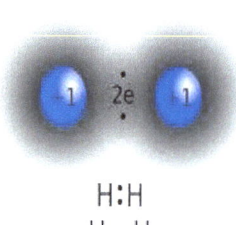

Bonds are of four types:

1. Ionic Bond
2. Covalent Bond
3. Metallic Bond
4. Coordinate Bond

Ionic Bond

When metal reacts with nonmetal they undergo complete transfer of electrons. Metal has a tendency to lose the electrons while nonmetal has a tendency to gain electrons. When metal losses its electrons it forms positive ion while when nonmetal gains electrons it forms negative ions. These ions attract each other with strong electrostatics forces to form ionic bond. For e.g. NaCl has ionic bonds extending in three dimensional space to form a lattice. Ionic lattices are closely packed arrangements which accounts for high rigidity of ionic compounds.

IONIC BONDING

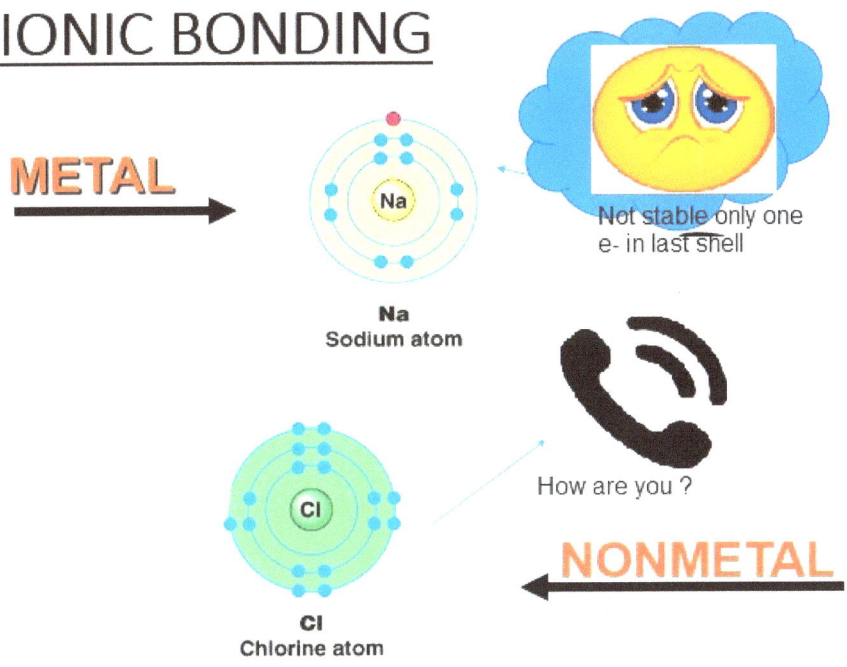

Illustration above shows how chlorine and sodium come in contact to form ionic bond in sodium chloride.

Sodium and chlorine react together; sodium gives its electron to chlorine. Now both elements have a full outer shell, but with a charge (Na^+, Cl^-). Now they are ions.

The two ions have opposite charges, so they attract each other. The force of attraction between them is strong. It is called an ionic bond.

When sodium reacts with chlorine, many sodium and chloride ions form and they attract each other. But the ions don't stay in pairs. They cluster together so that each ion is surrounded by 6 ions of opposite charges. The pattern grows until a giant structure of ions called lattice is formed. The overall charge of the structure is 0 since 1 positive charge and 1 negative charge neutralize each other.

Example : Formation of Sodium chloride

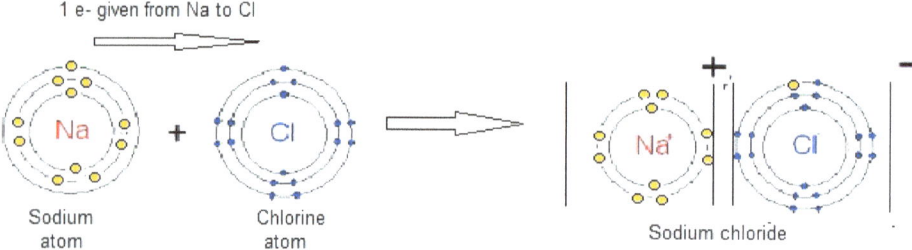

Lewis Diagram for NaCl

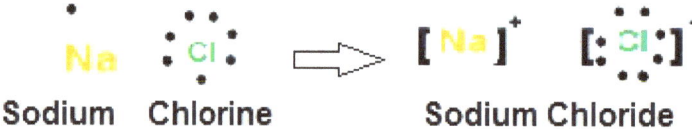

Other examples of Ionic bonding are K_2O, MgO, CaF_2 etc.

Lattice : It is a regular arrangement of cations and anions in three dimensional space of an ionic compound.

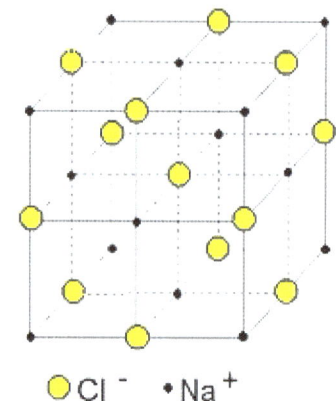

Lattice Structure of NaCl

Properties of Ionic Compound

1. Ionic compounds have high melting and boiling points.

This is because ionic bonds are very strong, so it takes a lot of heat energy to break up the lattice.

2. Ionic compounds are generally soluble in water. The water molecules can attract the ions away from the lattice. The ions can then move freely, surrounded by water molecules.

3. Ionic compounds can conduct electricity in aqueous or molten state. When melted the lattice breaks up and the ions are free to move. Since they are charged, this means they can conduct electricity. The solutions of ionic compounds conduct electricity too because they are also free to move.

Covalent bond

This type of bond is formed when two or more nonmetals mutually share electrons to form a bond (molecule). Thus molecules are covalent in nature. When a pair of electrons is shared, it is called a single covalent bond, or just single bond.

When 2 pairs of electrons are shared, it is called a double covalent bond, or just double bond.

Example : Covalent bonding in HCl molecule is shown below.

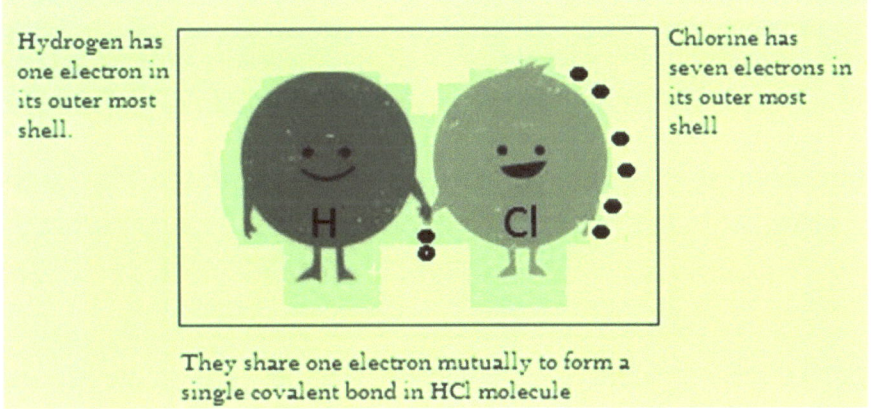

Example -2 : Bonding in NH_3

In ammonia nitrogen has five electrons in last shell therefore it share one electron each with three hydrogen atoms involving only three of its electrons.

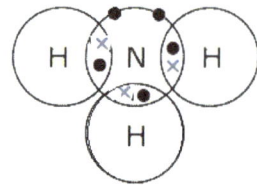

Other examples of covalent bonding are

CH₄, H₂O etc.

Molecular substances/Giant Covalent Structures:

Most simple molecular (covalent) substances are gases or liquids at room temperature. Molecular solids are held in a lattice but the forces between the molecules are weak. All molecular solids have similar structure. The molecules are held in regular pattern in a lattice. So the solids are crystalline.

Properties of Covalent Compounds

1. They exist in all three states of matter.

2. Covalent compounds have low melting and boiling point. This is because the forces between the molecules are weak.

3. They are generally insoluble in water but soluble in organic liquid.

2. They do not conduct electricity except graphite. This is because molecules are not charged

Giant covalent structure:

A giant covalent structure, or macromolecules are made of many of atoms bonded together by extensive covalent bonding such that structure extends in all directions in three dimensional space.

For e.g. Allotropes of carbon such as graphite and diamond are giant covalent structures. ***Allotropes are naturally occurring form of an element that have different physical properties but similar chemical properties.***

Diamond is made of carbon atoms held in a strong lattice. Each carbon atom forms a covalent bond to four other carbon atoms forming a tetrahedral rigid structure.

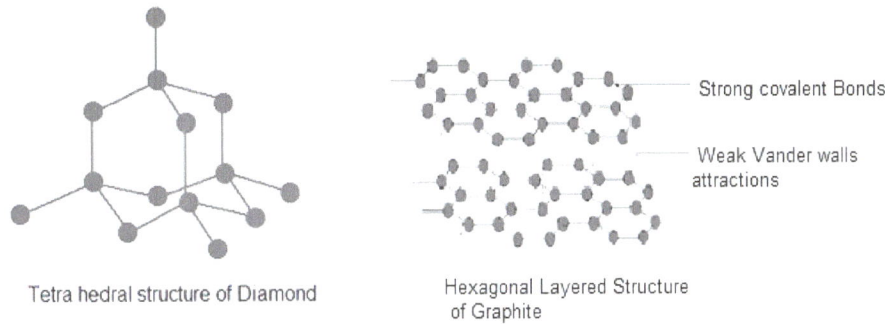

Tetra hedral structure of Diamond

Hexagonal Layered Structure of Graphite

Properties of Diamond

1. It is very hard because each atom is held by four strong bonds to other carbon atoms.

2. It has a very high melting point because of the strong bonds.

3. It can't conduct electricity because there are no free electrons to carry the charge.

Uses of Diamond

1. In making jwellery.

2. In cutting of heavy metals.

3. In drilling of earth.

Graphite

In graphite, each carbon atom forms a covalent bond to three other carbon atoms forming a hexagonal layered structure. The layers are stacked one over the other by weak bonds.

1. Is soft and slippery because the layers can slide over each other

2. Is a good conductor of electricity because each carbon atom has four

valence electrons and graphite is only trivalent so the fourth electron is free to move carrying a charge.

Uses of Graphite

1. In making pencil lead

2. In making lubricants

3. In making electrodes.

Silica or Silicon Dioxide

Silica is a giant macromolecular structure in which each silicon atom is covalently bonded to four oxygen atoms. Each oxygen atom is covalently bonded to two silicon atoms. Thus the formula is SiO_2.

Silica is a very **hard substance**. It has a **high melting and boiling point**, is insoluble in water, and does not conduct electricity. These properties result from the very strong covalent bonds that hold the silicon and oxygen atoms in the three- dimensional giant covalent structure. Silicon dioxide is found as quartz is naturally found in granite, sandstone and beach sand.

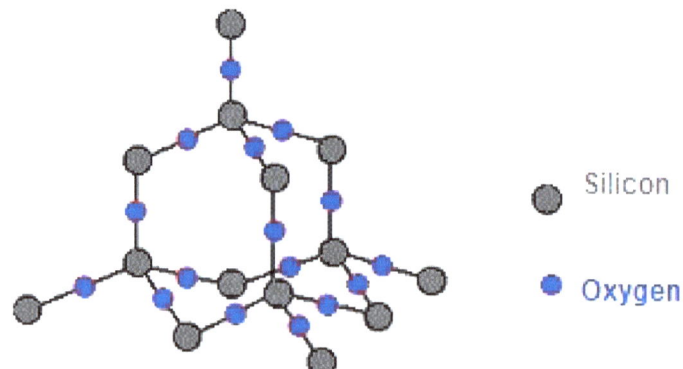

Uses of Silica

1. As an abrasive in making sand paper
2. In making ceramics and walls of furnaces.

Metallic Bond

Metals are electropositive in nature therefore their atoms loose electrons and form cations. These delocalized electrons form a mobile sea of electrons and are free to move. The metallic bond is the strong force of electrostatic attractions between this mobile sea of electrons and positively charged metal ions (kernels). Metallic bonds are strong, and layered as the layers can slide over each other so metals can maintain a regular structure and usually have high melting and boiling points.

1. *Metals have high melting and boiling points*

This is because it takes a lot of heat energy to break up the lattice of strong force between metal ion and electrons.

2. *Metals are malleable and ductile.*

Malleable: They can be beaten into sheets. Ductile: They can be drawn out into wires. This is because the layers can slide without the metallic bond breaking, because the electrons are free to move too. Below illustration shows a piece of copper metal (lattice) before and after hammering.

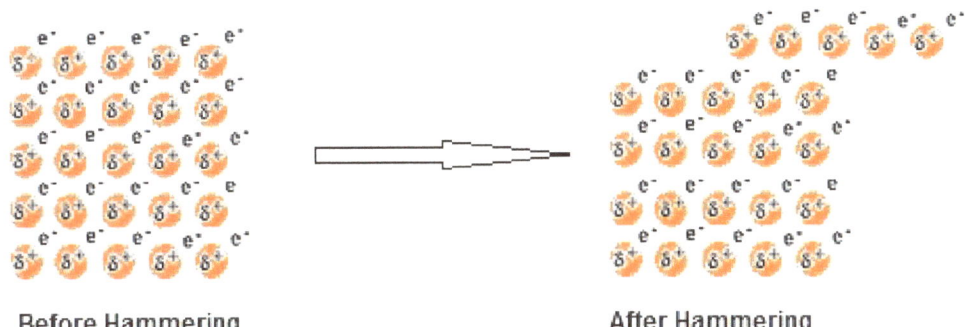

Before Hammering — After Hammering

3. *Metals are good conductors of heat*

That's because the free electrons take in heat energy, which makes them move faster and they quickly transfer the heat through the metal structure.

4. Metals are good conductors of electricity

This is because the free electrons can move through the metallic lattice carrying the charge.

Coordinate Bond

This type of bonding is found in compounds containing non-metals with lone pair of electrons. It is bond formed when both the electrons of a shared pair between two atoms belong to a single element. Thus sharing is not mutual rather one sided. Though this type of bonding is beyond IGCSE syllabus. The compounds that form these kind of bonds are generally Lewis bases that have lone pairs of electrons.

Example NH_4^+

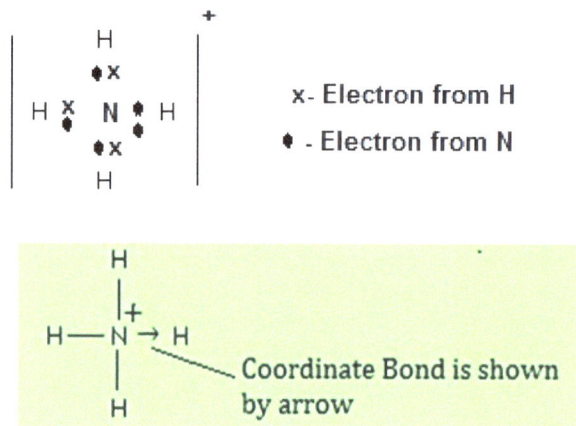

Valencies of common cations

Ion	Formula	Ion	Formula
Sodium	Na^+	Aluminum	Al^{3+}
Potassium	K^+	Copper II	Cu^{2+}
Calcium	Ca^{2+}	Iron II	Fe^{2+}
Magnesium	Mg^{2+}	Silver	Ag^+

Barium	Ba^{2+}	Nickel II	Ni^{2+}
Zinc	Zn^{2+}	Lead	Pb^{2+}
Rubidium	Rb^+	Nitonium	NO_2^+
Ammonium	NH_4^+	Tin II	Sn^{2+}
Hydronium	H_3O^+	Cobalt III	Co^{3+}

Valencies of common anions

Fluoride	F^-	Carbonate	CO_3^{2-}
Chloride	Cl^-	Bicarbonate	HCO_3^-
Bromide	Br^-	Phosphate	PO_4^{3-}
Iodide	I^-	Phosphide	P^{3-}
Oxide	O^{2-}	Phosphite	PO_3^{3-}
Sulphide	S^{2-}	Nitride	N^{3-}
Sulphate	SO_4^{2-}	Nitrate	NO_3^-
Sulphite	SO_3^{2-}	Cyanide	CN^-
Hydroxide	OH^-	Dichromate	$Cr_2O_7^{2-}$
Acetate	CH_3COO^-	Permanganate	MnO_4^-

How to write a formula with above valencies :

Example -1: Sodium sulphate

Write the ions from the above tables- Na^{1+} & SO_4^{2-}

Now exchange the valencies (without +/- signs) such that **Na_2SO_4**.

(Note - One is not written in the formula as SO_4 in above means one SO_4)

Example -2: Barium phosphate

Write the ions from the valency tables - Ba^{2+} & PO_4^{3-}

Now exchange the valencies (without +/- signs) such that **$Ba_3(PO_4)_2$**.

Example -3: Ammonium carbonate

Write the ions from the valency tables - NH_4^+ & CO_3^{2-}

Now exchange the valencies (without +/- signs) such that **$(NH_4)_2CO_3$**.

Technical Terms:

1. Bonds

2. Ionic, covalent

3. Metallic bond

4. Strong and weak bonds

5. Allotropes

6. Giant covalent structures

Exam Style Questions

Section - A: MCQ's

1) A metal X forms sulphides with the formulae XS and X_2S_3. Where is X in the Periodic Table?

A in Group III
B in Group I
C the 3rd Period
D in the transition elements

2) Which change to an atom occurs when it forms a positive ion?

A It gains an electron.
B It gains a proton.
C It loses an electron.
D It loses a proton.

3) For which compound is the formula correct?

A Magnesium nitrate - $MgNO_3$
B Calcium chloride - $CaCl$
C Potassium Sulphide - K_3S
D Silver nitrate - $AgNO_3$

4) The below figure shows bonding in the molecules of three substances.

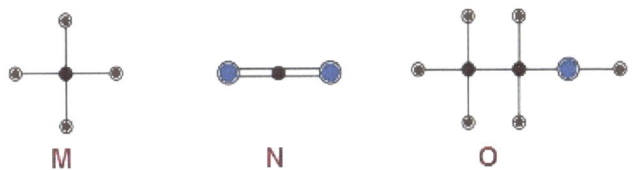

Which of these substances can be represented as below?

A M is NH_3
B N is H_2O
C O is CO_2
D N is CO_2

5) Which of the below numbers are added to give the mass number of a cation?

A number of electrons + number of neutrons + number of positive charges

B number of neutrons + number of protons + number of positive charges

C number of electrons + number of protons + number of neutrons

D number of protons + number of neutrons

6) In the compounds NH_3, SiH_4 and Al_2O_3 which atoms use all of their outer shell electrons in bonding?

A N and Si

B N and H

C H and Si

D Al and N

7) What happens to an atom that forms (2+) ion?

A. It gains two protons

B It gains two neutrons.

C It loses two electrons.

D It gains two electrons.

8) For which radicals is the formula correct?

A Sulphite- SO_3^{2-}

B Sulphide- S^{3-}

C Nitrate- NO_3^+

D Carbonate - CO_3^-

9) The diagrams show some substances in 4 different containers

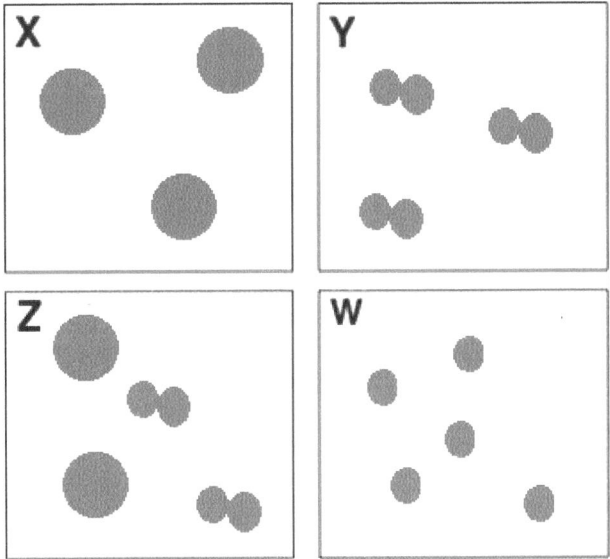

Which of the statements is correct for container Z

A it contains pure sample of Helium

B it contains a mixture of helium and hydrogen

C it contains a mixture of Krypton and hydrogen

D it contains only water vapor

10) The diagram shows an excerpt of part of the Periodic Table

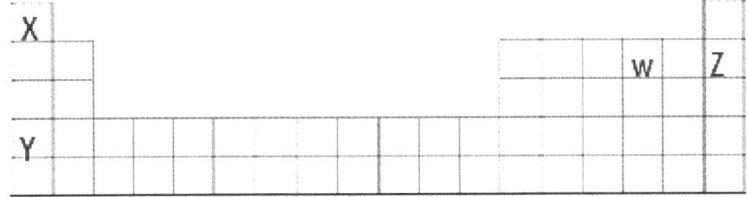

Which two elements could have a metallic bond in them ?

A W and Z   **B** X and Y   **C** X and Z   **D** X and W

Section B: Structured Questions

1. Methane thiol is a natural substance found in the blood and brain of humans and other animals as well as plant tissues. It has formula CH_3SH

(a) Draw a diagram that shows the arrangement of the valency electrons in one molecule of this covalent compound. Use **x** to represent an electron from a carbon atom. Use **o** to represent an electron from a hydrogen atom. Use **+** to represent an electron from a sulfur atom.

b) Predict two properties of this compound on the basis of its bonding.

....

....

2) Barium -Ba, is a metal that forms an ionic chloride $BaCl_2$. Iodine- I, is a non-metal that forms a covalent chloride ICl_3.

a) Which compound is likely to have the higher melting point and why?

....

....

b) Which of the above is expected to be soluble in water and why.

....

....

c) Rubidium oxide is an ionic compound. Draw a diagram that shows the formula of the compound, then charges on the ions and gives the arrangement of the valency electrons around the negative ion. The electron distribution of a rubidium atom is 2,8,18,.8,1

Use o to represent an electron from a rubidium atom.

Use x to represent an electron from a chlorine atom

d) Describe by the help of a simple diagram of rubidium chloride, what is ionic lattice. The lattice is same as sodium chloride lattice.

e) The reactions of these metals with oxygen are exothermic.

4K(s)   +   O$_2$(g) → 2K$_2$O(s)

(i) Give examples of bond forming and bond breaking in this reaction.

....

....

(ii) Explain using the idea of bond breaking and forming why this reaction is exothermic.

....

....

....

3) The electron distribution of a Sulphur atom is 2 + 8 + 6.

(i) Sulphur forms an ionic compound with potassium. Draw a diagram which shows the formula of this ionic compound, the charges on the ions and the arrangement of the valency electrons around the ions. Use x to represent an electron from an atom of potassium. Use o to represent an electron from an atom of sulphur.

(ii) Draw a diagram showing the arrangement of the valency electrons in one molecule of the covalent compound sulphur dichloride.

(iii) Predict two differences in the physical properties of these two compounds.

....

....

4) Sodium reacts with nitrogen to form the ionic compound, sodium nitride.

(i) State the formula of the sodium ion. ....

(ii) Deduce the formula of the oxide ion. ....

(iii) In the above ionic compound, the ions are held together in a lattice. Explain the term lattice.

....

....

(iv) What is the ratio of sodium ions to nitride ions in the lattice of sodium nitride? Give a reason for your answer.

.... sodium ions : .... nitride ions

....

....

5)-a) Iron is a transition metal. Describe the bonding in iron metal.

....

....

....

....

b) Based on the type of bonding explain why iron is a good conductor of electricity.

....

....

c) Properties like conductivity are important indication of type of bonding in a compound. Silica's conductivity was tested as follows

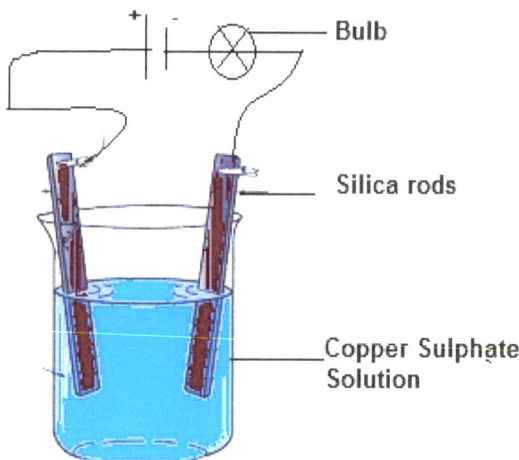

i) Explain why in the above setup the bulb does not light?

....

....

ii) Explain what will be the change in observation if silica rods are replaced by graphite rods

....

....

iii) Explain why the bulb may light brightly if silver rods are used in the above setup in place of silica rods.

....

....

....

d) Diamond is also covalent structure however it is very hard as compared to graphite. Explain this difference on the basis of bonding in these two substances.

....

....

....

....

e) Write two uses of silica

....

....

f) Another covalent compound used in making herbicides is methyl amine CH_3-NH_2. Herbicide are sprayed as shown in fields.

Draw a dot and cross diagram(valence shells only) showing structure of methyl amine if all atoms have single bonds.

****

www.ingramcontent.com/pod-product-compliance
Lightning Source LLC
Chambersburg PA
CBHW051019180526
45172CB00002B/402